MINER'S DAY

PARTHIAN

Parthian, Cardigan SA43 1ED
www.parthianbooks.com

First published in 2021
ISBN 978-1-913640-38-5
The Modern Wales series receives
support from the Rhys Davies Trust
Designed by Olwen Fowler
Printed by Gwasg Gomer
Published with the financial support
of the Books Council of Wales
British Library Cataloguing in Publication Data
A cataloguing record for this book is available
from the British Library.

Main text © Estate of B. L. Coombes
All images © see p. 161
Introduction and glossary © Peter Wakelin
Series Editor: Dai Smith

B. L. COOMBES

MINER'S DAY

Isabel Alexander
Rhondda images

Edited by
Peter Wakelin

CONTENTS

	page
Introduction	7
Miner's Day	46
Glossary	160
Acknowledgements	161
Index	162

Condemned Houses, Blaencwm, 1943, Watercolour and crayon on brown paper, 38 x 42

This painting may have been done on parcel paper, reflecting the difficulty for Isabel of finding or affording materials during the war. Although the houses that Isabel drew at Lower and Upper Terrace, Blaencwm, had been condemned, she noted that only the roofless house was uninhabited. She showed a figure coming from them.

Pithead and Slagheaps, Trealaw, c. 1943–5, pencil and watercolour, 31 × 27

This unfinished drawing shows Pandy and Anthony Pits at Naval Colliery, between Penygraig and Trealaw, with Llwynypia in the background. Acid-green grass is spreading over the tips, which are slumping into the Rhondda river. Children are using the exposed slopes as a slide.

INTRODUCTION

E. M. Forster's epigraph for *Howards End*, 'Only connect', is one of the most potent exhortations in twentieth-century literature. Published in 1910, the novel charted an unlikely friendship between a low-paid clerk and the cultured and well-off Schlegel sisters. Within the narrative Margaret Schlegel restates Forster's message: 'Connect – connect without bitterness until all men are brothers.' The coalminer and writer B. L. Coombes and the artist Isabel Alexander were among many left-leaning creative people who sought to foster communication across the class divide in the decades that followed. Coombes said as late as 1947: 'There is still a very deep pit between the mass of workers in this country and the intellectuals. It is that pit that we, the working-class writers, are trying to bridge.'[1]

Coombes and Alexander each had a passion to show the real life of coal-mining communities to a wide public, Coombes through words and Alexander through pictures. Coal mining was veiled more opaquely than almost any other sector from the view of educated people and metropolitan elites. For most of the country (at least until after nationalisation), the population on the coalfields was isolated in geography and distant in condition. Many of its particularities, literally hidden below ground, were unimaginable to outsiders who had never contemplated what it might be like to spend the day in darkness, to labour in a space no higher than a dining chair, to bear the daily crush of fear that a husband or father might not come home from work, or to suffer injustices of wage cuts, industrial disease and injury. Yet this invisible sector was not an obscure irrelevance. It dominated large tracts of the country, directly employed 766,000 people in 1939 and fuelled every part of Britain and its Empire.[2] In Wales, coal at its peak employed a quarter of the adult male population.

Against this divided and divisive culture Coombes and Alexander – Bert and Isabel – with other similarly engaged contemporaries believed the reality of mining communities had to be placed within the imaginative reach of far more people, and especially those in powerful positions. Atrocious working conditions and the scandalous deprivation and strife experienced by miners and their families had to be removed for good, but change would only come through understanding. Although they were concerned immediately and specifically with South Wales, much of what they had to say had wider application. It echoed coalfield communities in Appalachia, Asturias, Silesia, South Yorkshire or other regions.

The two of them came from very different backgrounds – Bert a self-educated miner who had grown up in poverty and Isabel a trained artist and qualified teacher. They were joined together only in their collaboration by way of text and images, which was published by Penguin Books in December 1945 as *Miners Day*, 'By B. L. Coombes / Illustrated by Isabel Alexander'.

Bert wrote the text over several months through 1944 and early 1945. It would be his fifth and final published book in a life's work as a direct witness that made an outstanding contribution to public understanding of coal-mining communities. Isabel provided images to interleave with the text while single-mindedly undertaking her own remarkable project from 1943 to 1945 to record people and places in and around the Rhondda valleys. Her nearly one hundred surviving images are a unique addition to the visual representation of mining communities in South Wales, although virtually unknown in Wales and never fully represented until now.

The book was a compromised production. Though heralded as a 'Penguin Special' it was a standard paperback of 128 pages and considerably slimmer than it should have been owing to wartime standards of economy and paper shortages. The eleven chapters were densely set and the quickly yellowing paper was so thin that it was hard to know whether you had turned one leaf or two. Only four full-page portraits and five vignettes could be included, and they were grainy and grey even though Isabel had made lithographs specially to unify and strengthen them. The title, too, was an oddity, and made no sense. Why was there no apostrophe in *Miners Day*? (Perhaps it was the designer's preference not to interrupt the Gill Sans capitals.)

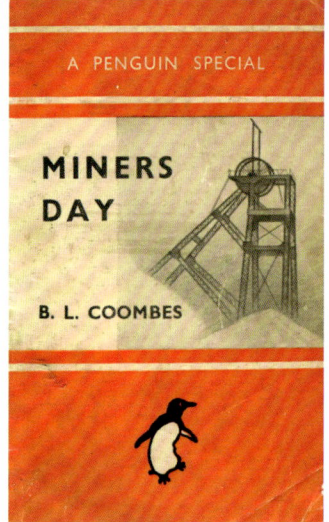 

This edition, under the title *Miner's Day*, is a chance to give Bert's still-significant text room to breathe and to present virtually all the images from Isabel's South Wales project that can be found. Most have never been reproduced before. A glossary, index and introduction have been added.

■ ■ ■

'B. L. Coombes' was the chosen name of the rather more exotically christened Bertie Louis Coombes Griffiths. His life and work are set out by Bill Jones and Chris Williams in their study for the Writers of Wales series as well as in his two volumes of autobiography. Born in Wolverhampton in 1893, he attended elementary school at the centre of the South Wales coalfield, where his father was a collier at Treharris. The family then moved to

Herefordshire, taking the tenancy of a small farm at Madley, between Hereford and Hay-on-Wye, in about 1905. Bert hated the bleak poverty and solitude of the farm and the stagnant social stratification of rural society. At seventeen he decided like thousands of other young men of the era to quit the countryside for the industrial valleys. Some forty miles away and at its zenith of production and employment, the South Wales coalfield held out the possibility of new adventures, bright lights, people, the comfort of coal fires and opportunities to earn hard cash.

He chose the upper Neath valley (he claimed he put places into a hat and picked one out at random) and arrived in 1910. He would spend more than forty years there as a miner and live close by for the rest of his life. The valley was different in character to the Taff, where he had lived as a child, or the nearby Rhondda, where Isabel would base herself. The local mines were mostly worked by drifts into the hillside rather than shafts. Much of the coal produced was the hard, jewel-like anthracite favoured for enclosed stoves, making it an area of growing employment even in the 1930s when the steam-coal producing pits of the central coalfield were declining. The area was also far less urban than the Rhondda, its scattered mining villages interspersed with green meadows on the valley floor.

Soon after arriving, Bert met Mary Rogers, who came from a Welsh-speaking family and was the daughter of the local lodge secretary for the South Wales Miners Federation. They married in 1913. Their first rented home was a damp, two-up-two-down terraced house abutting the New Inn in Mary's home village of Resolven.

After some prosperous years together the industry hit the profound industrial strife of the 1920s. It began with pay-cuts and with the lockout and three-month strike of 1921, which used up all their savings. The extended strike that followed in 1926 left them struggling to feed and clothe their two children. Inevitably, this was a formative time intellectually for Bert and put him on the first steps of his path to being a writer: it simultaneously forced him to confront the predicament of miners and their families and gave him time to widen and deepen his reading.

From the late 1920s, Bert worked at the Empire colliery, a series of drifts and air shafts that drilled a mile into the mountainside on the south-east flank of the Neath valley. He was variously a hewer at the coalface, the operator of a coal-cutting machine, a repairer and an ambulance-man, but always underground. He habitually took the night shift. This lifestyle of days above ground and nights below gives extra resonance to the title *Miners Day*: it suggests people and experiences beyond the colliery and his own work.

The Empire colliery and its environs were the book's setting. The villages of Cwmgwrach and Blaengwrach melded with one another, towered over by a conical waste tip on the riverbank for Aberpergwm colliery on the other side. Wooded hillsides rose to meet the bleak mountain heights of Craig-y-Llyn, above the head of the Rhondda. The place had the components that characterised most mining settlements: terraced houses of various distinctions, pubs, chapels, and a welfare hall and recreation ground provided by the Miners' Welfare Committee. By the 1940s, a pithead

Neath Valley, c. 1943–5, Conté, 32 × 26

In the briefest of sketches, Isabel has captured a good deal of information at Bert's Empire Colliery. She shows a double-railed passing place on the tramway up the wooded side-valley of the Gwrach. A loaded tram, packed high with large lumps of coal, is on one line to go down and an empty is on the other to go back to the mine.

baths and canteen had been built next to the railway station, the colliery sidings and the coal screens (all now blasted away by the A465 dual carriageway). The larger village of Glyn-neath felt like part of the same community though it was a mile away, and a mile beyond that began the glorious 'waterfall country' in the foothills of the Brecon Beacons that had been made famous by Romantic artists and writers who were exploring them just as the valley canal and earliest mines were opening in the 1790s.

Shift workers came to the baths by bus, train, bicycle and foot to change into working clothes and then walk the steep, often slippery incline up the colliery tramway, past the edge of the village and through the twisting side-valley of the Nant Gwrach to the cluster of engine houses and offices around the entrance to the mine. Here, the mine slope dropped into the workings and they walked down, parting ways to their respective jobs: cutting coal at the advancing faces, operating underground machinery and haulage engines or, like Bert, maintaining roadways and repairing roofs. The Empire mine, like many collieries, waxed and waned. At its busiest it employed 800 men underground and 130 on the surface. Two of Isabel's sketches, presumably made on a visit to meet Bert in 1944, capture the tramway in its narrow valley and the houses crouched below the mountain (left and page 149).

Bert started to write at around the age of forty and was soon gripped. By writing he sought to raise the consciousness of his fellow workers as well as to counter the wider ignorance so detrimental to their lives. He explained that when two workmates had been killed by a roof fall,

'I realised that neither coroner, solicitors, or hardly any one present had the least idea of what happened underground'.³ But it is also clear that he shared the passion to create of many mining people that generated singers, actors, writers, thinkers of all kinds; some world-famous, others private. As the Rhondda-born writer Gwyn Thomas put it: 'there was something so volcanic about the social experience of the Rhondda and similar mining valleys that people had this absolute compulsion to be lyrical, to be expressive, to leave some kind of expression of what they felt upon this earth.'⁴

He wrote at the kitchen table amid the clamour of two children and brass-band practice in the pub next door. Having left school at twelve, he educated himself as a writer by examining his rejections and analysing how successful authors had achieved effects. He also joined some classes run by the National Council of Labour Colleges.

In 1935–6, an article was accepted by *Welsh Labour Outlook* and an essay by *Left Review* and he began work on a novel.⁵ He sent a short story titled 'The Flame' to the recently established, book-sized literary magazine *New Writing*.

A portrait of Bert Coombes with his grandson during a period of unemployment, by Bert Hardy, January 1941.

Its energetic editor John Lehmann was keen to find new talents from diverse backgrounds; in due course he published Laurie Lee, Alun Lewis, George Orwell and Stephen Spender among many others. Lehmann fell excitedly on Bert's contribution, struck by 'the simplicity and unforced, quiet movement of the writing' and published it in 1937 to much positive attention.[6] Bert began work on his autobiography, encouraged and advised by Lehmann, to whom he dedicated it.

The autobiography was the greatest success of Coombes' writing career. It was published in 1939 under the title *These Poor Hands: The Autobiography of a Miner Working in South Wales*. Brought out by Gollancz, it was guaranteed strong sales as a Left Book Club Choice but it exceeded even these expectations. It was praised in the national press, among others by Cyril Connolly, J. B. Priestley, V. S. Pritchett and the former miner D. R. Grenfell, who would be Secretary for Mines for the duration of the war. Reviewers saw qualities that would be maintained throughout Bert's writing, namely that it was modest, self-effacing and all the more persuasive for rejecting a campaigning voice to tell an honest and dignified story which revealed the struggle and injustice of mining life. Arthur Horner, Communist and President of the South Wales Miners' Federation (and after nationalisation to be Secretary of the National Union of Mineworkers), felt that it laid bare 'the price of coal' in human terms but fell short by giving little space to the work of the union. Nevertheless, he acknowledged it as persuasive in its bare truth, without the explicit outrage and argument that might have led some readers to dismiss it as propaganda: 'for the facts revealed will speak for themselves'.[7] The book sold 80,000 copies in its first year and was translated into several languages.[8] After a period out of print, new editions were published in 1974 and 2002.

Bert became prolific. His pamphlet *I Am a Miner* came out slightly before *These Poor Hands*. He crafted more short stories, outstanding among them 'Twenty Tons of Coal', in which he drew on the memory of seeing his two friends killed. It was published in *New Writing* in 1939 and has been anthologised many times since.[9] After the Second World War broke out Bert began a column for his local paper, *The Neath Guardian*, in which he expanded on such topics as Neath bus station, the absence of church bellringing, mining and farming life, the waterfalls of the Neath Valley and trips to the seaside at Porthcawl.[10] He adopted campaigning approaches to problems to be solved once war was over in articles for *Picture Post* and a pamphlet with the Liberal peer Lord Meston, *The Life We Want*. An historical and technical outline of the coal industry for a general audience was published in 1944 as *Those Clouded Hills*. Bert had found a range of registers as a writer that suited his personality and interests. He kept up with other authors who had been translating the same themes for wider audiences at around the same time: he admired the realism of unemployed miner and novelist Lewis Jones's *Cwmardy* (1937) and *We Live* (1939), but (like most people close to the reality of the coalfields) disdained as fantasy Richard Llewellyn's international bestseller *How Green Was My Valley* (1939).[11]

The documentary photographer Bert Hardy visited the Coombes family to produce a photo essay published in *Picture Post* in January 1941. Mary Coombes is seen working at the coal range while caring for her three grandsons.

Bert and Mary had moved from Resolven to a cottage nearer to Cwmgwrach in 1938. A period of unemployment followed, as documented in a photo essay for *Picture Post* in 1941 that showed Bert taking time with his grandchildren and walking the hills, but thanks to *These Poor Hands* the family were able to rent a smallholding called Ynys-gron on the hillside looking north over Glyn-neath. At this point Bert could have put mining aside to write and farm but he returned to working underground. It might have been for an abundance of caution about his family's livelihood, yet he undoubtedly felt he could 'connect' as a writer only if he was – as it were – still at the coalface; he would feel disjoined from new realities, deracinated, if like many writers of working-class beginnings he moved on.

The stylised portrait of him that Isabel drew as the frontispiece of *Miners Day* (page 14) shows a determined stare, strong jaw and virile mane of hair to match his work ethic and social commitment, though it contrasts with the

reticent, kindly face one meets in *Picture Post*. The grandfather seems much younger than his fifty years (perhaps Isabel worked from a photograph taken a few years earlier). His collarless shirt and loose jacket emphasise his credentials as a working man. His backdrop is composed from his environment – the river Neath, wagons waiting to be filled, the sharp-edged triangle of the tip on the valley floor and behind it the steel stack of Aberpergwm colliery's power station.

Bert's practice was to use a notebook to capture thoughts and incidents underground. After the night shifts he spent much of his days running the smallholding with Mary and the two children or fulfilling wider commitments as a St John Ambulance man, a committee member of the Miners' Rest Home at Porthcawl and union Lodge Secretary.[12] He might be thinking and composing while he went to and from the mine or laboured in the fields and seams, but he had to quarry out time to sit and write. He said: 'Against me was almost everything – heavy, exhausting toil every night in the pit, lack of room, for my writing had to be done on the kitchen table'.[13]

■ ■ ■

B. L. Coombes, 1944, Conté, 44 × 32

Isabel's youthful portrait of Bert Coombes with a landscape of the Neath valley behind him that appeared in the opening chapter of *Miners Day*.

At the time of her South Wales project, the challenges faced by Isabel Alexander were not so different, despite the apparent security of her background. Her husband had recently left her, unsupported, six weeks after the birth of their son Robin. She was struggling to make ends meet, taking what work she could find and living in a shabby London bedsit. (Her parents were unable to help financially, having had to sell their house to pay for the treatment of her middle brother, Richard, in a psychiatric institution.)[14]

Isabel was born in 1910 and grew up in Birmingham, where her father, Joseph Manton, was a headmaster. She did well at school and played hockey for the England youth team but she was clear that she wanted a career in art. Her ambition was to study at the Slade School of Art in London, famous for having produced generations of great painters, among them Gwen and Augustus John, Percy Wyndham Lewis, Dora Carrington, Paul Nash and Stanley Spencer. It was the only art school of the time at which women were permitted to attend life-drawing classes. Nevertheless, her father would not countenance her going to London alone as a young woman of nineteen so she instead spent four years at the art school in Birmingham. After this she left home and taught for a year so that she could afford to move to London, where she took a further course in drawing and painting at the Slade. She won a prize for life-drawing and her early surviving works confirm that by now she had an effortless facility for drawing and a sympathy stylistically for Post-Impressionism and Vorticism.

After the Slade she remained in London, took a post at Bromley County School for Girls and in the lead-up to the war found opportunities to visit galleries in France, the Netherlands and Germany. Remarkably, she saw the infamous exhibition of confiscated 'degenerate art' or '*Entartete Kunst*' put on by the Nazis in Munich in 1937 and a huge exhibition of Vincent van Gogh at the Kröller-Müller Museum in the Netherlands. In London she saw the 1936 International Surrealist Exhibition and the display of Picasso's *Guernica* in 1938. These gallery visits opened up rare opportunities for insights into what were, in British terms, radical approaches, including social realism and expressionism. In Germany she probably encountered examples of the movement that would be most akin to her Rhondda project, New Objectivity or *Neue Sachlichkeit*. This was an approach to both painting and photography developed in reaction against the extremes of expressionism and abstraction. 'Objective' artists did not remake the world according to their own vision but recorded it. Thus, painters such as George Grosz and Otto Dix (both included in the 1937 Munich exhibition) depicted everyday things and people from all social classes soberly and with factual precision, often against symbolically encoded backgrounds. Isabel may also have known the widely published 'objective' photo essays of Edith Tudor Hart, who had trained at the

Isabel Alexander in around 1945

Bauhaus and lived in the Rhondda in the mid-1930s, where she took photographs as witness to injustice.[16]

In London Isabel got to know Donald Alexander, a friend at Cambridge of her younger brother, Guy. When they married in 1939, Isabel gave up her teaching job, as was expected of women at the time. Donald must have raised Isabel's awareness of the coal communities, since he was established already as one of Britain's pioneering filmmakers with a particular interest in them. He had made a documentary in the Rhondda with fellow students in 1935, where his association with local contacts in the Communist Party attracted police attention (he himself joined the Party after the war).[17] His footage of the unemployed scrambling up tips to scavenge coal became famous in Ruby Grierson and Ralph Bond's *Today We Live* (1937) and he returned to film in South Wales in 1937 while working for the documentary company Strand Films and its Director of Productions Paul Rotha. Later he set up the Documentary and Allied Technicians' Co-operative (DATA) and made films as head of the National Coal Board Film Unit.[18]

When Donald left Isabel in 1941, Paul Rotha (who coincidentally had preceded her by a few years as a student at the Slade) helped her get back on her feet by offering her periodic work as an art director that could be fitted around her care for her son. It was Rotha who suggested that she visit the Rhondda to see what might come from it artistically.[19] The Rhondda Fach and Rhondda Fawr valleys had been the epicentre of the Victorian steam coal boom and were packed with some 160,000 people before their terrible decline set in after the First World War. Local unemployment rates persisted at from 25 to 75 per cent and resulted in debilitating poverty and malnutrition. The economic upswing of the Second World War had brought measurable improvements but the area was still ingrained with the consequences of poverty and the agitation in the form of strikes, hunger marches, demonstrations and political debate that labelled it 'Red Rhondda'.

Isabel's first period of concentrated drawing was over several months in 1943. She returned for extended visits in the next two years. Her parents looked after Robin while she was away. It is extraordinary that she committed

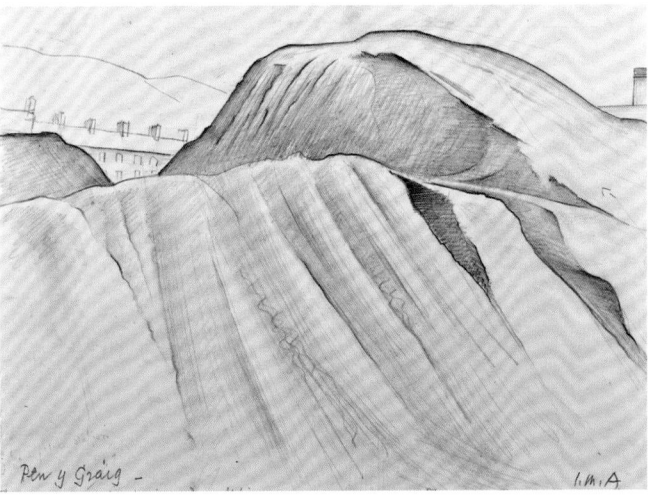

Penygraig, c. 1943–5, pencil, 34 × 22

The waste tip of Naval Colliery's Anthony Pit in the middle of Penygraig rises up towards the community of Tylecelyn Road. The chimney of the pit is on the right. Isabel wrote 'from police station' on the border and the viewpoint was close to Grovefield House on Dinas Road, which became a police station in 1920.

Blaencwm, 1943, pencil, 20 × 23

Glenrhondda Colliery at Blaencwm was remote from the valley settlements at some 800 feet above sea level with its own housing for miners' families. Beynon Row was crammed in below the Hendrewen pithead between a track, a railway bridge and the Selsig brook. The whole site was later cleared.

Isabel Alexander, London, 1938

herself to such a demanding and apparently unfunded project while she was struggling financially. Donald put her in touch with Jim Morton of Tylorstown, an unemployed miner and Communist Party councillor on Rhondda Urban District Council with whom he had stayed in 1935. Morton and his wife took in lodgers at 26 and 27 Vivian Street in Tylorstown and Isabel probably stayed with them: she made many drawings around Tylorstown and one of Vivian Street itself, apparently from their front garden (page 144).[20]

Isabel's three known periods in Wales took her to several locations judging by her surviving work. She produced many drawings around Blaencwm, at the very head of the Rhondda Fawr, of people, condemned houses and Glenrhondda colliery, which employed 480 men in 1945 and incorporated Blaencwm level, where she went underground.[21] It was extraordinary that she obtained permission to draw below ground as an outsider and as a woman, especially under the additional constraints of war. As only three brief sketches are known (pages 51, 70 and

Dinas from Trealaw Zigzag Path, c. 1943–5, pencil, 20 × 26

The long-demolished buildings in this drawing were four-storeys tall at Dinas Road and River Row. Isabel noted that the lower two storeys were separate houses built into the hillside: they would have had no light from behind and no through-ventilation.

101), it is likely she was not allowed to repeat the visits to do concentrated work, but the experience must have underpinned her comprehension of her whole project.

Other paintings and drawings show subjects clustered around the middle Rhondda at Dinas and neighbouring Trealaw and Penygraig. She probably went by bicycle to Cwmparc, up a side valley a few miles along the Rhondda Fawr, and across the mountain to Gilfach Goch, which was only two miles away as the crow flies but would have been a tortuous journey by train. In preparation for *Miners Day* she spent a week at Cwmgwrach around the end of March in 1944 and Bert rushed her around the area. She sketched the valley mines as well as portraits of Bert and two or more colliers.

Penygraig, c. 1943–5, pencil, 22 × 32

The tips at Penygraig seem like waves of a tsunami crashing up the slope towards the streets. Yet more tips from Naval Colliery are poised on the hill behind (also pictured on pages 6, 16, 19, 29, 94, 110, 132).

Untitled, c. 1943–5, pencil, 20 × 25

Isabel liked to strike up the mountainsides to capture the amphitheatre of the mining valleys. This view looks across the Rhondda from above Trealaw at the tips on the hillside behind Penygraig and Tonypandy.

The ventilation fan house and its chimney was all that remained of the former Nantgwyn Colliery. On the right, the diagonal of the tramway incline took spoil from Naval Colliery to tip on the mountain.

Stylistically, it is significant that Isabel did not follow the neo-romantic fashion that beguiled British art in the 1930s and 1940s. Her images of the coalfield are realist in idiom as well as subject matter, without the touches of surrealism and abstraction that coloured the neo-romantic work as War Artists of Graham Sutherland at mines and steelworks or John Piper at bombed-out buildings. Nevertheless, Isabel's style did vary subtly as she considered and reconsidered her approach. Her initial sketches from life, whether portraits or landscapes, were observational and accurate. In many she recorded notes about colour, which suggests that she was contemplating making paintings from them. The drawings of people are unselfconscious but soulful, sensitive and sympathetic. For example, her drawing of a teenaged boy called Graham who had been evacuated from Swansea, perhaps after witnessing the devastating Swansea Blitz of 1941, captures all the detail of his school uniform and the fine knit of his balaclava but still conveys sadness and isolation in his long face and dark, downcast eyes.

Graham, Evacuee from Swansea, 1944, pencil, 36 × 27

Many children were evacuated from the cities to the Rhondda valleys during the Second World War. Graham may have been sent away from Swansea after the Blitz of 1941, when 230 people were killed in three nights of bombing.

Blaencwm from the Mountain, 1943–4, watercolour, 45 x 50

This view of Glenrhondda Colliery at Blaencwm – nicknamed the Hook and Eye for its unusual headframe – was the prime image in the essay 'Coal: The National Plague Spot' in 1946 and was published, unattributed, in Coal, the NCB magazine, September 1947. It shows the higher of the two Glenrhondda shafts at the head of the valley with headframe, engine house, chimney and smiths' shops. The tips fan out from the pithead, bisected by the steep incline to take coal down to the screens and railway. The houses of Upper and Lower Terrace are on the far left. A few fields and gardens fight for the remaining space under the tips.

A few carefully rendered drawings of Rhondda streets adopted a hard-edged rigour to express the bleakness of the regimented, unembellished houses. Worked-up paintings such as her panorama of Blaencwm (above) relished formal qualities and found a kind of beauty in the composition without forsaking objectivity. Some pushed observed images further towards archetypal truths, as can be seen by comparing a watercolour study painted on the mountainside at Gilfach Goch with the finished picture published by Future Books over the caption 'The legacy of the coal rush', in which foreground interest has been introduced, details have been altered for pictorial effect and an eye-catching farmhouse has been removed (pages 30 and 22).

MINER'S DAY 21

Interesting contrasts also played out between the lithographs she made for reproduction and her preparatory drawings (pages 33 and 88): the portraits became more stylised as they were redrawn, broken subtly into fractured planes and often combined with backdrops that drew on an iconography of tips, terraces and mountains in the manner of New Objectivity. A handful of paintings took on a dreamlike quality, heightening the strange shadows around ruined houses (page 108), the ghostliness of figures (page 112) or the frightening approach of tips (opposite).

Her Rhondda project might in due course have led on to an exhibition but practical circumstances were against her.

The Legacy of the Coal Rush, c. 1944, gouache, 46 × 56

This image was given its title in the essay 'Coal: The National Plague Spot' in 1946. It was based on the study *Gilfach Goch from the Mountain* (page 30). Isabel adjusted the iconic image. She removed the distant farmhouse and some of the adjacent terraces, she made the tips more towering, and she added foreground grasses and filled out the haphazard fence (sketched in the earlier drawing) with angle-iron, corrugated sheets and salvaged wire from colliery cables.

Washday, 1944,
pastel, 45 × 36

The heroic struggle to put out clean washing among filthy coal tips, which Bert described, is illustrated by this image. It may be a composite impression of many places or based specifically on the houses at Blaencwm, which looked down towards the Rhondda valley.

After preparing the nine lithographs for *Miners Day* she provided ten paintings and drawings to be featured in 1946 in the lavishly produced, if idiosyncratic, *Future Books 1: Overture* to illustrate an essay titled 'Coal: The National Black Spot'. However, that was where the project came to an end. Subsumed in other work, she never returned to the same kind of social realism or to the South Wales valleys.[22] The Rhondda project remained essentially incomplete, though it comprised probably over a hundred paintings, lithographs, drawings and preliminary sketches, most of which she held onto throughout her life.

■ ■ ■

The collaboration seems to have been put together by the publisher, Penguin. By 1943 or 1944 Bert would have been looking towards a next book after the publication of *Those Clouded Hills* and Isabel had begun her visits to the Rhondda. The Penguin proprietor, Allen Lane, started the company in 1935 to provide accessibly-priced books in paperback for a mass market and his imprint Penguin Specials was designed to tackle topical issues from a centre-left perspective. Bert told John Lehmann that Penguin had commissioned a new mining book from him and had sent a 'lady artist' to sketch for illustrations. There was clearly a rapport - he described her to Lehmann as 'real good'. Did Isabel hear about the forthcoming book and contact Penguin about her work or did Paul Rotha tell Lane about her Rhondda project? The artist Barnett Freedman was another friend that Isabel and Allen Lane had in common.

Alternatively, it is possible that Penguin approached Isabel to commission her Puffin Picture Book *The Story of Plant Life*, which would finally be published in 1946, and that the illustrations for *Miners Day* were either a first test or an opportunistic extra. Whatever the route taken, Isabel's images made a striking contribution, marking the book out as Lane intended as something 'special' and contemporary.

Miners Day came out a few months after VE day and shortly before the Japanese surrender. The prolific flowering of attention and reviews that welcomed *These Poor Hands* was not repeated. If books have moments, Bert's autobiography had hit its moment in 1939 but the moment for *Miners Day* perhaps had passed. Since the 1930s a succession of books, photo essays and documentaries had made the coalfields less opaque to the rest of the country than it had been before. Added to this, the end of the war was so much of a disjuncture that something written before it and published as it ended might have seemed less relevant. Change for Britain and its coalfields had begun already under the wartime coalition and the reforming Labour government that followed it.

A notice in *The Listener* was sympathetic but brief. It described *Miners Day* as 'a series of glimpses of mining life, inside the pit and in the village', prompted by a desire 'to make the coal-user realise what sort of life is lived by the coal-getter.' It pointed to the long wait for change in the industry: 'It is this sense of avoidable frustration that Mr. Coombes so admirably conveys.'[23] Few other notices appeared. *Miners Day* was never reprinted and was removed from Penguin's lists in 1951.

Isabel's images made an important contribution to the book and these together with her larger body of work in the coalfield serve as fascinating reflections of a time and place. Since then, the valleys and their culture have changed beyond all recognition. Daily lives no longer answer to the rhythm of the mines and the land itself has been transformed through the reclamation of tips, the demolition of collieries, the contraction of townships, new development and greening. (The painter Charles Burton refers to the blackened Blaencwm of his youth now looking like the forests of Alsace.)

Isabel's art stands out as a special contribution to comprehension of the South Wales coalfield partly because she was embedded to a degree that few outsiders were. Her commitment to objectivity placed her on a fulcrum of social realism that described South Wales as it was, without toppling into either outrage or encomium. Isabel recorded what she saw with understanding of its implications but not the horror of visitors who flitted in and out of the coalfields seeking to awaken the nation's social conscience. In this sense her closest comparators are perhaps the photographer Edith Tudor Hart and the painter Cedric Morris (later a friend). She eschewed the polemical approaches of insider artists in the Depression who had idealised like Vincent Evans or railed like Archie Rhys Griffiths, and equally the mission to condemn poverty and hardship of visitors such as Maurice Sochachewsky or her former husband Donald Alexander. Yet the wartime interval in which she came was just before the more benign era of post-war advances when the insider generation of Charles Burton, Ernest Zobole and others in the Rhondda Group viewed their home environment with warm familiarity and the émigré expressionist Josef Herman could praise the miners of the Swansea valley as icons in his vision of humanity.

As to the text, *Miners Day* is not a book with a grand conception or driving themes. Perhaps Bert had emptied his store of those in his earlier books. The last chapter of *These Poor Hands* had represented a typical day and *Miners Day* proceeded to describe not one but a diversity of days, like strands woven to make the pattern of a mining community. Bert allowed himself to wander through the scenes and subjects that caught his attention. Each episode feels meandering and relaxed. He may even have thought to emulate the stream of consciousness approach of Virginia Woolf, with whom he was in literary dialogue in 1941.[24] It is as interesting to dip into as to read from start to finish. There are moments of lyricism, moments of passion, but as in all Bert's works, its chief qualities lie in its observational directness free of the pretention and formality that often bogged down non-fiction of the era.

■ ■ ■

The first miner's day opens – opposite that youthful portrait of the author, so determined in his purpose – with a soundscape in the darkness of the blackout as Bert walks up the incline to his night shift: the drone of a plane flying far above, the pant of the engines forcing compressed air below, a snarl of buses bringing people to work, the roar and whistle of a train, greetings of men passing downwards.

This is the daily repetition, 'our regular cycle as day follows day and shift meets shift'.

From the beginning, comradeship and craftsmanship express themselves as Bert's most essential themes. They intertwine as he and his team repair roofs and keep roadways open. Characters and their conversations seem to be observed, though he gives some people pseudonyms. The affectionately depicted 'George' demonstrates the nature of the craft most clearly as – a grocer until recently – everything is new to him and has to be explained. George doesn't function underground like those brought up to it and Bert knows he will be forever clumsy, without 'pit sense' [52]. Each skill, it seems, requires a 'knack' not easily explained. When George tries to shift a derailed tram he strains and grunts, exerting his full force without moving it an inch, but his slighter colleague Steve steps in

Dan, 1944, lithograph, 34 × 24

Isabel wrote on this lithograph, made for publication in the original edition of *Miners Day*: 'L. E., Totally Disabled from Silicosis', and on a photograph of it that she had drawn him in his own home. In the book it was titled 'Dan' and accompanied Bert's tender description of his friend who was dying from dust disease. It seems likely that 'Dan' was a pseudonym and 'L. E.' represented his real name. He probably sat for Isabel during her week in the Neath Valley. The original Conté drawing is in the Glynn Vivian Art Gallery.

26 MINER'S DAY

and flips the tram then swings it into its position [54]. Several stories about George among his mates inject comradely humour. When he thinks he knows better than to use the standard miner's food tin and water jack and instead wraps his sandwiches in a cloth and brings a thermos flask, the outcome is disastrous. He loses his food to rats and his tea to the rock, but his mates insist on sharing with him what they still have [152-3].

The geology almost becomes a character itself as Bert ponders the capriciousness of unstable strata and the mysterious variations between stifling heat and chilling cold, between coal that falls to the touch and coal that calls for dynamite [84]. The forces playing through the strata are clear from Isabel's drawings of old steel arches twisted like rope and Bert's descriptions of the hazards of unbolting them under tension and trying to get them out [61-2]. Through one night Bert and his mate Crush (a nickname true to the black humour of the pit) watch and listen to stones running from the roof, cracks opening, timbers snapping, and movements simultaneously from below and above as the mountain seems to want to close-up the voids the workers have so impudently made: 'Strong posts thick as a man's body and quite nine feet long were being forced down through the softer upper flooring. And as the roof came down the floor was pressed upwards. That nine-foot post began to look like seven feet' [152]. Such timbers hold up a slab in one of Isabel's sketches at Blaencwm (page 70).

With more than thirty years' experience behind him, Bert was keen to explain the changes in his craft, from crawling to undercut a shallow seam by hand with 'a terrifying sense of being crushed and smothered until you get used to the roof being tight against your back and your face not far off the floor' to the easier labour but appalling noise and dust of driving a coal-cutting machine [134-5].

■ ■ ■

A subject that hangs like a black cloud over Bert's text and Isabel's images alike is 'dust'. The word describes the coal waste that clogs water, land and air as well as the crippling disease that its inhalation causes. Pneumoconiosis, silicosis, 'black lung' or plainly 'dust' was more prevalent in South Wales than anywhere in Britain in the 1940s. It had become an epidemic since the introduction of pneumatic drills and coal-cutting machines, bringing disability and death to multitudes of previously healthy miners. It is not yet fully understood but Bert notes that even a young man driving a haulage engine far from the coal cutters is suffering, he presumes from dust thrown up by the moving cable (87).

Bert recounts the case of Jerry, who has had to give up work and in due course dies while those who should have helped him argue about whether 'it was a weak chest or bronchitis, or an overload of coal dust or the scheming of a malingerer' [58]. The unthinking vicar at the burial chants his customary 'dust to dust'. Another former miner, 'Dan', who is battling the compensation board, was drawn by Isabel. Her portrait notes his initials 'L. E.' and says she drew him in his home, 'totally disabled from silicosis' (opposite).[25] The image matches Bert's description: 'Dan has one of the most expressive faces I have seen. Yes, it has become haggard and that blue tinge shows on his cheeks, but there is always that inspired look about him which convinces you he is thinking of something which seems wonderful. He has a rich, tuneful voice and his eyes have the sparkle of an enthusiast' [59]. Dan died before the book was published.[26] On the back of another portrait drawing of 'C. D.' Isabel noted: 'totally disabled by silicosis. Colour of face uniform yellow' (far right). A finished image of him is known from a black-and-white photograph but she was so upset that she destroyed the original.

In the last four months of 1944 alone, Bert notes, in just the western region of the coalfield, 2,440 men were stopped from working underground because their doctors suspected the disease [127-8]. Its terrible impact is expressed by the story of Gwilym who dies while the Whit Monday parade is going on outside. He would have been marching himself once, but for five years he was unable to work. When he could no longer get up the stairs his wife slept in a chair in the kitchen to listen for his gasping. The causes of his illness were unproven, so he told his wife: "I know it's dust […] and they've cheated us while I've been alive. Make them pay for me when I'm dead" [147-8].

L. M., Partially Disabled by Silicosis, c. 1943–5, conté, 31 × 25

The collar and tie worn by this man might suggest he is trying to find other employment after becoming disabled by dust disease. Such work was scarce in the coalfield. On a photograph of the drawing Isabel noted 'Young miner. Member of drama group and choir. Fine tenor voice.'

CD, Totally Disabled by Silicosis, Trealaw, c. 1943–5, conté, 33 × 28

Isabel's notes on this drawing read: 'A bad case of silicosis. Colour of face uniform yellow.' She completed a painting from the drawing but destroyed it. 'C. D.' seems to be a relatively young man, but his face is drawn and lined. His checked scarf is folded tight to his throat.

Dust despoils the whole environment. After going for an X-ray of his lungs in Neath Bert comes back on the bus through tracts of fields and rhododendrons, green and scarlet. He walks the last part towards Cwmgwrach: 'A hundred yards along that road the taste of dust reminded me that the colliery screens were at work. My eyes were gritted and that acrid, bitter feel came to my palate […] It sort of holds your breathing and dries up your tongue' [158].' The caption for one of Isabel's drawings reads:

Untitled, c. 1943–5, pencil, 20 × 27

Naval Colliery throws its spoil towards Penygraig. The foreground houses, since demolished, were on Crabtree Road (now a back lane called Old Brithweunydd Road). When published in 1946 in the essay 'Coal: The National Plague Spot' the caption read: 'Slagheaps among the houses – coaldust in the air, on the curtains, on the kitchen table. To men and women with a passion for cleanliness, it means washing, cleaning, scrubbing; washing, cleaning … All over again.'

From Penygraig, Road to Dinas, c. 1943–5, pencil, 27 × 34

This drawing of the Rhondda Fawr below the road to Dinas from Penygraig shows a cast-iron sewer pipe along the riverbed. Isabel noted that the water was black with coal dust. It was common to see sheep wandering the highways.

'coaldust in the air, on the curtains, on the kitchen table. To men and women with a passion for cleanliness, it means washing, cleaning, scrubbing; washing, cleaning … all over again.' Her painting *Washday* (page 23) presents a house hemmed in by tips, perhaps at Blaencwm or Cwmparc. The tips erode into a grey slurry that is filling up the intervening gulches. On a drawing of Dinas she noted a common sight in the valleys of South Wales, that the river flowed black (right).

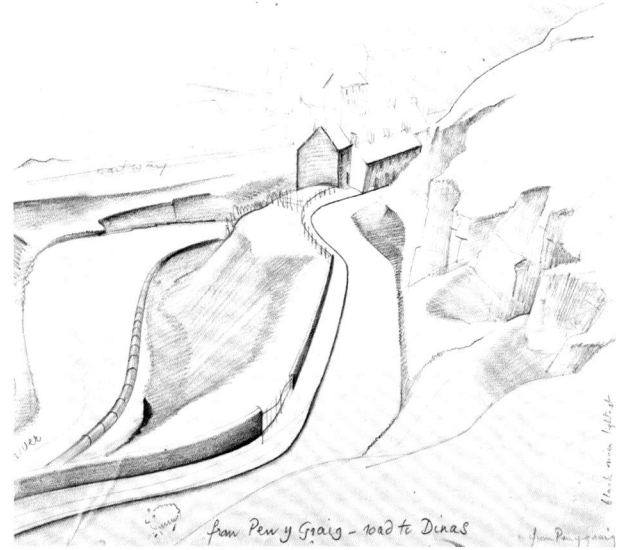

MINER'S DAY **29**

Gilfach from the Mountain,
c. 1943–5, watercolour, 27 × 38

This study for the finished image in the article, 'Coal: The National Plague Spot' (page 22) looks across Gilfach Goch from the High Street to Evanstown, the two communities dominated and divided by the massive tips in the valley.

Looking down over Gilfach Goch from the east she captured the straggle of houses along the High Street, the black mass in the centre of the valley and the terraces on the other side (opposite). Nothing seems to escape the black excrescence in a valley jammed with tips, pitheads, coke ovens and railways for the Britannic Merthyr, Glamorgan, Gilfach Goch and Glynogwr collieries, the river culverted beneath it all. At Tylorstown, the looming tip appears in several drawings, still active and expanding like the one that fills the valley at Aberpergwm, which Bert describes: 'All day an automatic tipper climbs steadily upwards towards the top of the black pyramid. On the high peak the tipper seems to hesitate slightly as if ashamed, and then over goes another few tons and the tipper crawls downwards again. The dust cloud lifts and consorts with the wind' [114].

■ ■ ■

In the first critical overview of Coombes' work, in 1974, Dai Smith pointed out that Bert was keener to describe the technicalities of extracting coal than the intricate life of the mining townships.[27] He was anxious in all his

Gilfach from the Mountain,
c. 1943–5, watercolour, 20 x 23

This sketch shows Gilfach Goch when the whole settlement was dominated by waste tips. It looks from the mountain behind High Street across a railway bridge in the valley to Gilfach Goch Colliery. The landscape has since changed beyond recognition.

MINER'S DAY **31**

writings to convey the hidden characteristics of work underground, but *Miners Day* together with its illustrations opened a wider view than his earlier books. Isabel injected subjects from the Rhondda and Bert, perhaps stung by Arthur Horner's view of *These Poor Hands*, expanded his themes and told others' stories alongside his own. This time he wrote of the iniquities of pay, the solidarity of South Wales miners, housing, entertainment and the openness of mining communities to incomers among many other topics.

Pay is a constant matter of contention. The miners may now be working to daily rates rather than paid by tonnage but the wage is low and the complexity of deductions and allowances is open to abuse. Deductions are made for the pithead baths, the doctor, the Miners' Federation, coal delivered to the house and income tax among other headings – a pay docket for April 1944 was reproduced to prove his point (page 129). Arguing with officials about time worked or deductions becomes an occupation in itself. Collieries were under state control for the duration of the war and a pay award had been negotiated by Lord Porter that resulted in miners getting five pounds a week at a time when the average for manual workers was six pounds and a shilling. Bert recounts the ensuing strike: 'Possibly our action was wrong in view of the circumstances, but it was but the climax to a series of happenings which should have been avoided. It was the outlet of a seething disgust for the continual delays and evasions, intensified by the feeling that men who knew nothing of our work or ideas were making decisions that would affect our lives and our families.' The breakdown leads eventually to a revised national agreement, heralding new industrial relations in a forward-looking post-war culture [63-5].

The 'Fed' and the solidarity of its members is the miners' main defence.[28] Bert records the scepticism of his friends Steve and Benjy about pedantic meetings and official slowness but he knows that 'whatever they may say about their union, they will obey a decision properly taken, no matter what the cost. Starvation, victimisation, or jail, whatever the threat, their loyalty to their mates in a time of crisis would transcend them all' [81]. However, he observes that eighteen years on from the 1926 strike the few who were blacklegs still get preferential treatment [106]. The company reveals its institutional vindictiveness when settling a brief strike at Cwmgwrach, which the miners called after it threatened to remove the right to cheap coal from men off work with industrial diseases. Months afterwards, the company is deducting money from their wages in pursuit of damages [134].

The impact of work on health is not just pneumoconiosis. Bert sees a group of injured men queueing at pit top for compensation pay: 'They were a battered assembly with arms in slings, crutches helping their legs, and often bandaged faces' [89]. He sees his friend Tom Evans, a collier for most of his life, with his two brothers who are a shopkeeper and a curate: 'Tom looked almost old enough to be their father. His frame was drained of every ounce of fat, his hands were curved inwards like claws, and his shoulders had a stoop. Of course he had plenty of blue marks on his hands and face' [90]. Isabel drew Tom when she visited the mine at the end of a shift.[29]

Her first, naturalistic drawing includes a swift note of his thin face and a deft portrait (right). At a time when helmets were not yet standard or compulsory a hat with a strip of metal riveted across the peak and crown sits battered and askew across his head. The lithograph she produced for *Miners Day* (page 88) was heightened and took on a tragic tone, literally darker, more sculptural and expressive, head and helmet now as one, suggesting not so much a man as an industrial machine. She notes he suffers from nystagmus, the disease of flickering eyes deemed to be caused by working in poor light.[30] Like Dan, he died before *Miners Day* was published.

The gains in conditions of recent years do not go unnoticed, such as the arrival of the welfare hall and recreation ground and the revolution of the pithead baths provided by the Miners' Welfare Committee where there are facilities to wash and change, a canteen, and buses to go home on. Bert recalls his earlier days of walking back from work in freezing weather wearing sodden clothes, and an occasion when he sat down with tiredness and only by luck was saved from hypothermia [81]. He contrasts Cwmgwrach with his former village, Resolven, where

Drawing for Tom, c. 1943–5, conté, 37 × 28

Isabel noted on the back of the lithograph (page 88) done from this drawing that she sketched Tom at the end of his shift. Bert wrote that Tom Evans was ageing faster than his brothers, who had other occupations. His face and hands had scars that were blue with ingrained coal dust. Few colliers wore helmets; this type was made from compressed paper.

Trealaw, 1943, pencil, 25 × 20

This is a view down Cairo Street in Trealaw from the corner of Pergwm Street. The grid of Edwardian terraces with back yards represents a definite improvement on the condemned houses cheek-by-jowl with pits at Blaencwm.

Blaencwm Condemned Houses (Inhabited), c. 1943–5, pencil, 20 × 28

Isabel noted that this row of dormer cottages at Blaencwm had been condemned but was still inhabited (it may have been the rear of Upper Terrace). A narrow strip of vegetable plots was tight against the waste tips with roughly-made fences to keep out the sheep.

no pithead baths has yet been built and its lack infects the whole community with dismalness [139]. Isabel's paintings and drawings capture landmarks and environmental textures that have gone for ever – headgear and winding houses, laundry poles, fragmented corrugated iron, fences made of colliery wire and acres of black tips.

Housing problems were alike in the Neath valley and the Rhondda. Isabel's drawings and paintings provide a remarkable assessment of the worst pre-war housing stock in the Rhondda. They constitute an important record given that many of these places were not recorded in any other way, disappearing while standards rose and the population

Colliery Wire,
c. 1943–5, pencil, 21 × 28

This drawing is titled 'colliery wire' because the fence is made of wire strands unravelled from winding ropes. The location may be Blaenllechau near Ferndale. Isabel noted: 'Whole thing's varying tones of indigo except for, bright light on roofs, dull ochre sides of houses and distant grass – and some bits of bright green in near grass. Sharpest contrast between fence and background, next between roofs and their surroundings.'

of the mining valleys tumbled. The houses that she drew and painted at Blaencwm were already condemned and partially vacated. Others occupied inhumane locations like the row called Tramway Side that crouched beneath the pithead at the top of Gilfach Goch. She showed a valiant attempt to carve a space for growing vegetables outside a condemned row at Blaencwm and allotments worked between old tips at Penygraig. The best houses, like the terraced, urban streets she drew at Trealaw and Tylorstown, had rear yards at least.

Bert's text weaves narratives around these visual records. He calls for better houses than the grim terraces thrown

Gilfach Goch, c. 1943–5, watercolour, 28 × 35

Britannic Merthyr colliery stood at the very head of the valley at Gilfach Goch, spoil tips rising around it. A bridge over the railway took waste to a further tip. The row of houses known as Tramway Side sat below the pithead and the bridge, tight alongside the railway. It employed around 600 men in the 1940s and closed in 1960.

up in a hurry by colliery companies that had resources to do better. He sees families housed in run-down army huts, the inability of young couples to rent better than two rooms. But he also sees models for improvement: houses offered by building clubs and semi-detached designs that give space and light. He describes a young woman in tears of excitement and relief because after ten years of huddling six to a room she had received the offer of a council house, with bathroom. Her letter, in an instant, 'altered her vision of life for herself and her four children' [141].

Even in the 1940s, many families were struggling. Bert sees a woman who is distraught at how to keep her children now that her husband, removed from mining because of early pneumoconiosis, is having to work in an

English factory and cannot send home enough for them to live on [95-7]. The elderly have difficulties too, out of work, dependent on their children for support: 'Sharing a common kitchen, and perhaps a bedroom, there is no privacy nor any quiet resting place' [113].[31] The strength of the community has been eroded by the introduction of three shifts. The afternoon and evening shift 'cuts workers off from everything. For them no evening walks, no cinema, no games and no social life in any form. This three-shift system has almost spoiled all the choirs, the bands and the dramatic societies which used to be so energetic' [115].

Isabel's portraits complement this inside knowledge. It is hard to think of any other painter or photographer who made so many portraits of coal-mining families in the period: the elderly, young miners, mothers and children, all treated as distinct individuals. She was not trapped by the conventional portrayal of colliers black from the pit but looked instead at the men underneath the mask of dust, those who had retired through injury and the women who kept families alive in desperate times.

In the 1940s there was still little paid work in the coalfield available to women but they were fully occupied by the challenges of managing on low incomes, looking after children and fighting the battle against the black muck that crept into the house with 'soap, scrubbing-brush and water, and then brown paper over the oilcloth so that it shall not be soiled' [112]. The frequent scrubbing of front doorsteps and pavements, as shown in Isabel's drawing on page 144, was symbolic of family pride. On the morning of Whit-Monday, a rare day of expected leisure: 'Men, ordered out of the way by busy housewives, lounged along the streets and collected in groups on the points where the street-ends came out bluntly against the main road. All along that street and all the other streets the wives were on their knees scrubbing the lone flagstone which fronted their own hole in the wall'. Simultaneously, they were keeping fires alight, making a special dinner and taking part in celebrations [144]. Mothers in the coalfield were known to starve themselves so their children might eat well enough. The strain shows in Isabel's portraits.

The originals of these portraits are now missing. *Mother and Child* shows an emaciated mother carrying a baby in a traditional shawl over her shoulder. *His Wife has Carried on Her Never-Ending Battle* was published in *Miners Day*, the title from Bert's text.

Pit Boy, Blaencwm, c. 1943–5, pencil, 26 × 19

Isabel wrote 'Pit Boy' on this drawing though its subject looks about twenty. He might have been a conscripted Bevin Boy or another young employee not yet recognised as a miner in his own right. She drew him at the pithead, presumably before he went underground at Glenrhondda Colliery.

Royston, Blaencwm, c. 1943–5, conté, 32 × 26

Most or all of Isabel's portraits of children seem to have been done at Blaencwm. She must have worked quickly to capture her sitters in a few still moments.

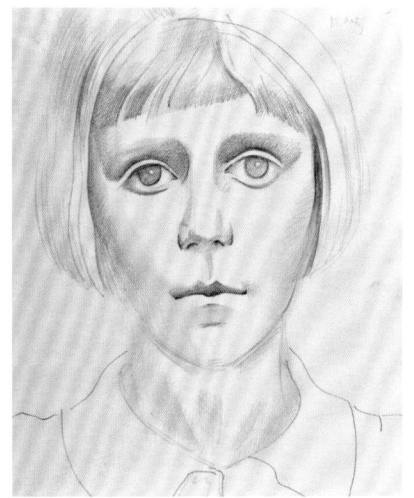

Margaret (left) and her unnamed younger sister, both c. 1943-5, pencil, 24 × 20

Terry, Blaencwm,
1944, pencil, 30 × 22

Isabel saw the children she drew with an objective eye as well as humane appreciation of each individual, neither sentimental nor caricaturing. She captures Terry's worried, downward gaze, his layered woollens and his pudding-bowl haircut.

Joan, c. 1943–5, pencil, 25 × 20

The drawing of serious, dark-eyed teenager Joan stands out as one of the most beautiful of Isabel's portraits of children.

MINER'S DAY

The one published in *Miners Day* (page 37) has the title 'His wife has carried on her never-ending battle'. Others show women at different points of their lives, different stages of despair. 'The women wrap their shawls tight around their babies and shelter them with their love,' Bert says, and Isabel shows the figure of a woman walking among pigeon lofts and washing poles with her shawl tied diagonally to hold a baby to her chest (page 67) and, close-up, an emaciated young woman clutching her baby (page 37). An older woman, 'Mrs G' may be the finest of all Isabel's lithographs (page 109): she is portrayed with deep humanity and recognition yet the wear of years of struggle show in her face and her unfocussed eyes. A caption says she has brought up seven children, her husband is an ex-miner and her sons and grandson still work down the pit.

Perhaps the most remarkable strand of Isabel's project was her study of children. Several pictures pick out distant gangs of children playing on the tips or scavenging coal to carry home but the series of portraits that she drew are an extraordinary rarity. A comparison could be made with the work of documentary photographers of the 1940s and 1950s in slum areas or with Joan Eardley's post-war paintings of kids on the streets of Glasgow, but where their images were emotive and often shocking, Isabel captured

January, 1944, pencil, 19 × 13

This footwear is evidence of the poverty in the Rhondda that persisted into wartime. The boy is wearing once smart boots now torn and split across both toes.

Children in the Streets, January, 1944, pencil, 19 × 11

While most of Isabel's sketches of children's legs show signs of poverty, this boy has a coat, a decent pair of old boots and woollen socks.

her subject's likenesses with interest in them as individuals. It showed how familiar she became to the communities she visited that she persuaded so many children to sit for her. Although we have no names for most of them, one can believe one sees their varied characters: Royston, bright and curious; Margaret, anxious and uncertain; Terry, resentful, wrapped in layers for warmth; Joan with the beauty Mrs G might once have had (page 39). One series of sketches showed the actuality of hardship and poverty in the depths of January 1944 (probably some 700 feet above sea level at Blaencwm): legs bare to the elements, feet inadequately shod, one child with evidence of rickets (page 127). The war years were a period of relative prosperity and better diet thanks to war work and rationing, but the legacy of the hungry thirties persisted in the damage done to minds and bodies. Bert speaks of his hope that excellent new schools and children's gifts for finding something good in dire circumstances will allow the coming generation to flourish.

Strangely, the war is little mentioned in the book, but a constant subtext is that raising coal is crucial to the war effort. There was, after all, heroism in common between miner and soldier – both jobs were perilous, physical, reliant on comrades and might bring injury or death. Bert was sceptical how much the arrival of the Bevin boys would contribute (chapter 7), though one of those 47,000 drafted workers, Michael Edmonds, in his own memoir saw their contribution far more positively (he too barely mentioned the war).[32] One moment most connects Bert's presence in the valley with the war so far away. He feels excitement and foreboding when he and his family sense the preparations for the invasion of Europe, the beginning of the end: '[We] saw a sight which checked our breathing. The half-darkened sky was starred with moving lights, red, blue and seeming yellow, which came towards and over us to the accompaniment of that unceasing flying roar' [126].

■ ■ ■

As the war came to a close and *Miners Day* was published, Isabel was fully occupied by a contract to write and illustrate *The Story of Plant Life*, which presented botanical concepts to children with clarity and graphic beauty. It was published in a large print run in 1946. A French version followed in the *Collection du Vieux Chamois*. She went on to research, write and vividly illustrate a sequel, only to be told in 1948 that Penguin had shelved it along with other intended titles for the Puffin Picture Books. It was never published, but an edition is currently forthcoming.[33]

For the first time Isabel gained some financial security in 1949 when she began a permanent post as a lecturer at Saffron Walden Teacher Training College. Living at Thaxted, she got to know East Anglian artists such as John Aldridge and Edward Bawden as well as the Welsh painter Cedric Morris who had stayed with mining families and painted in the South Wales valleys in the 1930s. Her continuing development as an artist was set out in 2017 in *Isabel Alexander: Artist and Illustrator* by Janet McKenzie and in a retrospective exhibition at the Mercer Gallery in Harrogate. After a period of more neo-romantic landscape painting in the 1950s she became absorbed increasingly by her own richly-coloured form of abstraction from nature. She travelled to paint again, as she had done by coming to the Rhondda, visiting Italy, Spain, France and many parts of Britain. She painted obsessively in the Hebrides, where she visited almost every inhabited island, and in Yorkshire, where she made her home after her retirement from teaching in 1977. As many women artists of her generation found, opportunities did not come easily and she was largely overlooked in her lifetime. Nevertheless, she participated in group exhibitions and had frequent solo shows from the late 1950s onwards in Cambridge galleries and Yorkshire. In 1995, she returned to her Rhondda material to plan an exhibition at Doncaster Museum and Art Gallery in the South Yorkshire coalfield, but before she could bring it to fruition she was taken ill. She died in 1996.

Bert did not publish another book after *Miners Day*. He wrote his column for the *Neath Guardian* and occasional articles or stories for magazines, including the National Coal Board's house-magazine *Coal*. He also attempted a second volume of autobiography and a second novel, but both were unpublished. He and Mary moved in 1945 to a larger farm near Onllwyn, high on the hills between the Swansea and Neath valleys, where they worked at winning a livelihood from raising stock and running a milk round. For the first time they bought a property of their own, using author's fees and royalties equivalent to several years of a miner's pay.[34] Always interested in output, whether coal or text, he estimated in 1947 that he had written 2,000,000 words. Finally, he left the mine after an injury in around 1955 that resulted in having to wear permanently what was called a 'steel waistcoat'.

Since Bert's death in 1974 a significant amount of his output has been available, including the anthology *With Dust Still in His Throat* edited by Bill Jones and Chris Williams in 1999, which published extracts from his second autobiography and his second novel, short stories and diaries. His papers are preserved in the Richard Burton Archive at Swansea University. *Miners Day* has been, until now, the most important of his works that was out of print.

■ ■ ■

It was a quiet agony to Bert when his son Peter decided to take work at the colliery. He wrote in *Miners Day*: 'something was dying in my inside when I watched his light disappearing after we had parted at the division on the main drift […]. I wished I had touched his face or given him a last handshake – yet men, and boys, I suppose, are not good at emotional partings' [115]. In the last few pages he contemplates a farm abandoned to coal workings, now themselves abandoned too. He seems to be prefiguring the end for mining in his valley and silently welcoming it: 'Such a system does not deserve to endure – it cannot last' [158-9].

Bert Coombes, though, has lasted. This book adds to the substantial sum of his writing in print. He has been appreciated widely. A Canadian civil servant, David B. Brooks, was so impressed by *Miners Day* when he stumbled upon a copy in the 1970s that he published an article about it. He wrote: 'I know of no other book that offers both so personal an account and so perceptive a critique of conditions in the British coal mines of the 1940s, yet it is practically unknown, at least in North America.'[35] In 1974, the Spanish literary critic Ramón López Ortega in his survey of the British literature of workers in the Depression said that Coombes in *Miners Day* had 'brilliantly depicted the collective personality of the miner'.[36] Bill Jones and Chris Williams argue that his 'everlasting achievement is that he succeeds, arguably more effectively than any other comparable writer, in making a mass audience vividly aware of the sheer humanity as well as the inhumanity of the world of mining.'[37] The coal industry historian Ben Curtis remarked in 2020: 'His work remains the best and most vivid description of the lived experiences of miners and mining communities in mid-twentieth-century South Wales that I've ever read.'[38]

Both the contributors to *Miners Day* were exceptional in their determination to give a whole and rounded view of mining communities. How many observers of industrial life in the period even saw the retired, the children and the women, let alone gave them that concentrated attention that can be seen in Isabel's portraits? Bert made mining understandable yet he did not conceal complexity or contradiction: he unpicked the knotted absurdities of pay; he featured the proud, skilled worker and the youthful fool who spreads false rumours; he characterised the union's machinations as well as its achievements. This was truth as both artist and writer saw it, not fiction or polemic. While Bert's newspaper column advocated for nationalisation, protection against silicosis or the bringing of the arts to every valley, *Miners Day* was diagnosis rather than prescription.

What did they achieve? Did they connect the working-class life of the coalfields with the nation's opinion-formers and decision-makers? The nation certainly did change after 1945, thanks to the nationalisation of the coal industry in 1947, the creation of the National Health Service, the science applied to pneumoconiosis, and later the reclamation of those great tracts of waste. John Lehmann believed that Bert's writing 'may have had something to do with the great stirring of national conscience', and the immediacy of Isabel's images for *Miners Day* and for 'Coal: The National Plague Spot' may have reached to further audiences. Alongside the activists and agitators, the politicians, the medical researchers, the photographers, the film-makers, they played their part.

What was a call to action then may be a call to action now, as workplace regulation and employees' rights are matters of renewed relevance, as child poverty and even child malnutrition are realities again, as decent housing is beyond the reach of many, and as the coalfields continue to suffer by every measure of wellbeing from their long inheritance of deprivation. *Miners Day* is relevant and useful still.

The publication of a new edition of Bert Coombes' text and so many images from Isabel Alexander's Rhondda project introduces their achievements to new audiences. They connect now not just across the classes but the generations, to a world in which the labour, the strife and the black dust of the coal industry have gone. There was 'so much bitter history to forget', Bert said. And yet remembering is necessary too.

NOTES

1. Radio talk, 'The Working-Class Writer', 1947, in Patrick Hannan (ed.), 1988. *Wales on the Wireless: A Broadcasting Anthology*. Gomer. pp. 157–9.

2. William Ashworth, 1986. *The History of the British Coal Industry, volume 5, 1946–1982, the Nationalised Industry*, p. 3. Oxford.

3. Quoted in Bill Jones and Chris Williams, 1999, 2nd edn. 2017. *B. L. Coombes*. Cardiff: University of Wales Press, p. 8.

4. Radio talk, 'Rhondda', 1980, in Patrick Hannan (ed.), 1988. *Wales on the Wireless: A Broadcasting Anthology*. Llandysul: Gomer, pp. 48–50.

5. The novel, 'Castell Vale', was never finished but selected chapters are included in an anthology of Coombes' otherwise unpublished work: Bill Jones and Chris Williams (eds), 2014. *With Dust Still in His Throat*. Cardiff: University of Wales Press.

6. Quoted in Bill Jones and Chris Williams, 1999, 2nd edn. 2017. *B. L. Coombes*. Cardiff: University of Wales Press, p. 26.

7. Arthur Horner, 1939. 'These Hands Build the World', *Daily Worker*, 28 June 1939, p. 7 (I am grateful to Daryl Leeworthy for supplying a copy); Bill Jones and Chris Williams, 2002. Introduction to B. L. Coombes, *These Poor Hands*. Cardiff: University of Wales Press, p. xix.

8. Bill Jones and Chris Williams, 1999, 2nd edn. 2017. *B. L. Coombes*. Cardiff: University of Wales Press, p. 47.

9. It was published most recently in Dai Smith, ed, 2014. *Story: The Library of Wales Short Story Anthology, Volume 1.*, Cardigan: Parthian.

10. Christopher M. Baggs, 1990. 'A "War-Time Mirror to Welsh Life"? B. L. Coombes and the Neath Guardian'. *Morgannwg*, 34, pp. 78–93.

11. Ibid., p. 85.

12. Ibid., pp. 81–3.

13. Coombes, 1942 in *A Miner's Record I*, quoted in Christopher M. Baggs, 1990. 'A "War-Time Mirror to Welsh Life"? B. L. Coombes and the Neath Guardian'. *Morgannwg*, 34, p. 90.

14. Janet McKenzie, 2017. *Isabel Alexander: Artist and Illustrator*, Edinburgh: Isart Press, p. 33.

15. Isabel Alexander, 1996. 'Sources of XXth Century Art in Europe: Experiments and Influences', unpublished extracts from writings about major artists, p. 57.

16. Wolf Suschitsky, 1987. *Edith Tudor Hart: The Eye of Conscience*. London: Dirk Nishen Publishing.

17. Transcript of interview with Donald Alexander by Paul Marris for Trade Films, courtesy of Robin Alexander.

18. BFI information, www.screenonline.org.uk; Dave Berry, 'Obituary: Donald Alexander', *The Independent*, 2 August 1993.

19. McKenzie, 2017, p. 10.

20. Personal communication from Robin Alexander, 2020; personal communication from Daryl Leeworthy, 2020; forum.ferndale-wales.co.uk.

21. Glen Rhondda, welshcoalmines.co.uk.

22. With the artist Elizabeth Rivers, Isabel worked for several weeks in 1947 on Inis Mór, off the west coast of Ireland, where she drew cottagers, fishermen and crofts, though in a more interpretive idiom than she had used in the Rhondda. Examples appear in Janet McKenzie, 2017. *Isabel Alexander: Artist and Illustrator*. Edinburgh: Isart Press, pp. 60–64.

23. *The Listener*, 13 December 1945, p. 699. Coombes wrote to John Lehmann on 3 April 1944: 'Had a lady artist here for a week doing sketches for a mining

Penguin they have commissioned from me. She was real good and might have something to suit you. I thought her real New Writing style in work and thought … Penguins sent her to me and she has a lot of material now after my hurried rush around the area with her.' Quotation courtesy of Chris Williams, from John Lehmann Correspondence, Harry Ransom Center, University of Texas at Austin. In a letter of December 1945 to Jack Lindsay, Coombes further described Isabel as 'a real left winger'.

24 *Folios of New Writing*, 3, Spring 1941.

25 Copy drawing, Glynn Vivian Art Gallery. She wrote 'Portrait of L. E. Rhondda' but this may have been added long after the fact and referred to her Rhondda project rather than the subject's location. Coombes' letter to Jack Lindsay, 7 December 1945, suggests that she did draw his friend.

26 B. L. Coombes, letter to Jack Lindsay, 7 December 1945.

27 Smith, David, 1974. 'Underground Man: The work of B. L. Coombes, "miner writer"', *Anglo-Welsh Review*, 24, pp. 10–25.

28 For a history of the union see Hywel Francis and Dai Smith, 1998. *The Fed: A History of the South Wales Miners in the Twentieth Century*. Cardiff: University of Wales Press.

29 Note on back of lithograph, Glynn Vivian Art Gallery. Coombes, letter to Jack Lindsay, 7 December 1945, suggests that she did draw his friend.

30 Miners' nystagmus is no longer recognised as a medical condition though it was widely recognised in Britain before the advent of improved lighting underground. For a review see Fishman, Ronald S., 2006. 'Dark as a Dungeon: the Rise and Fall of Coal Miners' Nystagmus', *Archives of Ophthalmology*, 124, pp. 1637–44.

31 Coombes' value as a primary source for the elderly in mining communities is apparent in Ben Curtis & Steven Thompson, 2015. '"This is the country of premature old men": Ageing and Aged Miners in the South Wales Coalfield, c.1880–1947', *Cultural and Social History*, 12, pp. 587–606.

32 Michael Edmonds, 2013. *War Underground: Memoirs of a Bevin Boy in the South Wales Coalfield*, edited by Peter Wakelin. Newport: South Wales Record Society.

33 It is hoped that the book will be produced by the Penguin Collectors Society and Design for Today in 2022.

34 His royalties from *These Poor Hands* alone in its first six months of publication were some £640, compared with a wage of £5 a week; Barbara Prys-Williams, 2000. 'A Difficult Man, Your Coombes', *New Welsh Review*, 49, pp. 55–60.

35 David B. Brooks, 1979. 'Retrospective: Miners Day by B. L. Coombes'. *Appalachian Journal*, 6, pp. 311–13.

36 Ramón López Ortega, 1974. La crisis economica de 1929 y la novelistica de tema obrero en Gran Bretaña en los años treinta. Salamanca, Spain, Gráficas Europa, p. 19.

37 Bill Jones and Chris Williams, 1999, 2nd edn. 2017. *B. L. Coombes*. Cardiff: University of Wales Press, pp. 103–4.

38 Twitter @DrBCCurtis 12 January 2020.

Overleaf:

Behind the Screens, Penygraig, c. 1943, conté, 28 × 36

This screens for sorting coal building at Penygraig was presumably the one that served the Naval Colliery's Anthony and Pandy pits. Loaded coal trams were pulled up a series of ramps and tipped out onto the screens. Although it looks dilapidated it was probably a fairly recent construction, built with a steel frame infilled in brick and a concrete slurry tower.

CHAPTER ONE

A chilled darkness had followed the March evening. Two sounds disturbed the night when I crossed the slope towards my work, forcing the left side of each boot into the ground to prevent my slipping. High above, a heavy plane moved through the night sky with a stuttering drone, while so close that the ground seemed to be always quivering I heard the ceaseless pant, pant, pant of the huge compressor engines which were gulping in the sweet mountain air and forcing it through pipes into the colliery working, so that many engines there, as well as the conveyors and coal-cutters, would have power to do their work.

Men were flying through strange surroundings high above me, and I knew that down below my feet, in that underground world where I would soon be, other men were working in unnatural conditions so that life in our country could go on. In the valley the mining villages were invisible; conquered by the blackout. Buses snarled or whined as they brought miners from distant areas to their work. Sometimes a torch stabbed the darkness for a bright moment, then faded, but that instant of light pulled my sight round and made me lose my exact sense of direction. The staccato warning of a whistle preceded the rumbling of a workmen's train, already easing its speed because it was nearing the station. The steam came up to shroud me whilst I walked, making me think of washing day at home with the boiler bubbling over the fire. Benjy and Steve and Dave would be on that train with a couple of hundred more mates of mine. Five minutes later, when I was quite near the colliery premises, another train roared in, bringing some hundreds of miners from another area. Hermit and George, with the others, would be sitting on those bare dusty seats. Probably they would be bringing something with them. A puppy, or cabbage plants, or a book, or perhaps a newspaper cutting; for there is a constant interchange of things which actually interest us.

My mates were coming from all directions, like an army of invaders who had surrounded their objective, towards that hole in the ground. In one more hour they would have met and greeted their mates from the afternoon shift, and those men would be homeward bound by train, or bus, or on foot if they lived close. Such is our regular cycle as day follows day and shift meets shift with its continual use of man's energy and skill to rob these mountains of their treasure.

To us it was just another start after a weekend of resting. The return to a job which we have done so many times that we do not notice its strangeness … yet that is not quite correct. We always sense the change both in conditions

E. L., the Overman, Blaencwm, 1944, lithograph, 39 × 32

The caption to this portrait in 'Coal: The National Plague Spot', 1946, said: 'E. L. is an overman – a sort of foreman or manager's deputy. He started work at 11 and has been at it for 53 years. His group gets the worst of both worlds – the men tend to think of it as on the owners' side, yet the wages are little above a collier's. E. L., however, is well-liked – the men find him firm but fair.' Isabel noted that he worked at Blaencwm but she based the background on her drawings at Gilfach Goch.

and surroundings, even in our breathing and thinking. During the weekend our mole-like existence fades from close consciousness, but is revived when we start again in that pit life which is our second nature. When we have cleared our throats and washed the grime from our bodies we feel so different; as a good dinner drives away the hollowness of hunger. That is the queerest part of it all—how swiftly we forget the problems of the pit; possibly because they are so far hidden from daylight. I have to stir my memory: 'What was the work we did at the end of last Friday's shift?' or, 'Where did I last leave my tools?' Then we get back amongst it, and we adapt ourselves to a continual bend and watchfulness. Were I twelve months away from mining I know that the dust of underground workings would be as remote to me as the bogey fears of childhood. I know it; that way it has been with so many.

Meanwhile I am in it every day. In some ways repelled by what I see and feel, yet in other aspects enjoying the battle with the inside of the mountain; but most especially happy to be a good comrade for most of these men who are moving alongside me now. We are black figures moving through a black night towards a black pit.

About halfway up that tram incline, where we stumble over iron rollers and hook our toes under steel ropes, or bang into a loaded coal tram, we meet the last of the afternoon men on his way down. He has found his pipe and smokes a pleasant-smelling tobacco. He greets each shadowy group with a 'Good evening, boys.' Always I return the greeting after the others, so that I shall again hear his reply. His voice is deep and clear, with a sound of culture in it. His greeting and the aroma of his smoking linger with us while we stumble along, like the last reminder of a comfortable world.

After we have sidled past the rows of trams, walked through the bank of steam which is the panting breath of the pit engines, hidden our smokes for the morning, and exchanged our number checks for a lamp, we go underground. The weather affects our speed in this direction. When it is wet or cold we hurry inside, glad of the earth's shelter, but on those nights, when the moon shows high and clear in the sky, we linger to extend and savour those last moments, looking upwards and thinking what such a night must mean to the lives and romances of others. Then, slowly, we move inwards before the threat of an oil lamp carried by the overman.

Inside the entrance the refuge holes are crowded by youngsters, averaging about twenty years of age. They are always disputing, viciously it seems. Their language is brutally profane. Girls, film stars, miners' agents, politicians, are all brought inside that bawling discussion and are cast out besmirched as they pass on to condemn others of whom their knowledge and conception must be very slight. Everything and everybody outside their group receives the same verdict of being, profanely and emphatically, 'no blasted good.' They are just a section of our mining youth, not the largest section by any count, but their insolence and indifference to all discipline make them a problem in our work and our future. They linger until the overman has almost reached them, then move inwards unwillingly, disputing and swearing as they go.

The overman follows, knowing they will only go so far and so fast as he makes them. What mistake in environment and education has brought these young lads into this condition?

Above us steel arches support the roof like the ribs inside a giant whale. Our lamps reflect on the water which flows blackly outwards along the slimy gutters. The long underground roadway is clustered with the lights of walking men—who seem to be hundreds of shadows which talk to one another about what news the radio gave last or the papers misrepresented: for the disbelief of what they saw in the papers seems unanimous.

I overtake two men, both elderly and named David. One is slight and short, the other quite the reverse. Before each shift they meet, greeting one another with 'Hallo. You've come?' Then, on the walk to the colliery, and through that two-mile underground journey they go on together without exchanging one word. After the shift is ended they wait at an inside cabin, assure each other they have come again and resume that wordless journey homewards. In the canteen they take strict turn to fetch the two cups of tea and sit silently next to one another until their train is due in. Possibly both can see their life and their usefulness passing away and feel comfort from this company of another man who is similarly placed.

Far inside we pass under a large engine which is built above the mine roadway so that trams can run underneath. The man who drives this engine is deeply religious and therefore a butt for the thoughtless amongst the youngsters ... and many of the adults. I respect him, without following his teachings, because he fights to keep his convictions under difficult conditions. My feeling is that the mining world is surely the stoniest soil on which the seed of religion has ever been sown. Many of our workers feel there is something inhuman about a man who does not blaspheme at his work, although in their outside lives they act as decent folk. Even in the dread surroundings of a bad accident I have never heard a prayer offered or suggested. But to return to that engine driver.

He is a lay preacher and likes to prepare his sermons. Last week I was coming in late after explaining some trouble to the overman; one of my jobs is as a miners' committeeman. All under the earth was so still as I hurried along under that engine when suddenly a strong voice proclaimed from high above me in the darkness: 'Unless ye repent of your sins ye shall surely perish.' Not being at all prepared, nor knowing which sin was meant, I jumped upwards and forwards rapidly and my heart did a lot of fluttering until I realised it was the engine driver doing a bit of rehearsing when he thought no one was within hearing. Another time, walking a yard behind my mate at this spot, a fairly large stone dropped from the roof twenty feet above and fell between the two of us edgeways, slicing the back of my mate's coat in its fall. 'Missus'll have a jib about mending that,' he stated, when he examined the rip.

When I talk of stones do not imagine something that you can pick up one-handed to throw at anyone who makes faces at you. This one was about half a hundredweight. With us a stone may vary from a few pounds to several tons. The only classification I notice is when something about, say, ten tons, drops and the miner surveys it as he complains, 'Another blinkin' pebble to break. Would blacken your toenail if it dropped on it.'

When we reach our various districts the fireman is waiting to test the lamps. Benjy is there and has the news that Will won't be in because 'his mother's wife is dead.'

It takes the fireman some minutes to realise that the deceased is Will's mother-in-law. 'Jawch,' argues the affronted Benjy, 'it do mean the same thing, don't it?' Benjy is short, bald-headed, and has full cheeks polished like red jam apples. His speech is a confusion of Welsh and English. More than fifty years of age, he has spent all his working life underground, but is still a labourer. No ambition, so says Steve, our haulier, who has an impish sense of humour.

'Don't you worry,' counters the aggrieved Benjy. 'This kid will go through the world all right. You believe me now.'

At his age, still in the same home, doing the same work at the same wage as he earned when starting, it does not seem Benjy is going to travel far, but he knows all the tricks, absorbs all the male and female gossip, and considers himself of no small importance. As Steve says, Benjy has two ever-recurring questions whenever a workman approaches him in the mine. They are: 'Where is he?' (he being the fireman) and 'What time is it?'

A newcomer, who has worked with me since illness knocked out my regular mate, is waiting for me by the tools. I am interested in this new mate for two reasons.

Coal Face, Blaencwm, c. 1943, conté, 34 × 24

Isabel had an opportunity to sketch underground in one of the levels at Blaencwm, which was part of Glenrhondda Colliery. Three drawings are known to survive. This shows a miner – perhaps a repairer like Bert – silhouetted by his lamp under the broken roof of an arched roadway, so low that he has to stoop.

MINER'S DAY **51**

I like him very much, and I am interested in watching the effect of mining work and life on a man who came to it first in middle age. He is nearing forty, quite sixteen stone in weight and fairly intelligent. He has been underground only during the last few weeks and, in his own most emphatic wording, will not be there one day more after peace is declared. He ran a business of his own and was doing very well, but would have been taken into some branch of the Services if he had not elected to go into the mines instead. This way he tried to keep his business going during his spare time. He had been reared in a mining community and has many miners in his family. Compared to the ordinary shopkeeper he would, in the past, have been counted quite an authority on mining matters. In the future which George hopes will hurry along, he will again be able to speak with his weight on mining, but his outlook will be different—very different.

George is willing to work and learn, but he started too late. You can see it in his every move. He is too clumsy, his big body is set too firmly in a comfortable mould. In time the mine will melt every ounce of that fat from his bones and may make that frame a little more bendable, but George will always remain the man who struggles vainly to get a respectable amount on a shovel and to throw it to the exact point where he wants it; and it seems he will never learn how to time the blow of a mandrel or hit a stone with the sledge head instead of his body. He seems totally unable to learn pit sense. He wears a shiny safety helmet; he is the only one amongst us that does. As we move forward he misjudges the height of the roof and there is a sharp tang as the helmet hits the rock and then rolls on the floor. He bends ponderously, then replaces the helmet.

'If I hadn't had this helmet,' he states, 'I would have knocked half my head off before now.'

'No fear, you wouldn't,' I dispute with him. 'You would have learnt to bend far enough after the first couple of taps.'

The air current lifts behind us like a strong wind. We are going before it, so the dust caused by our walking clouds away before us as we travel. Small coal is shoe deep on the roadways, lumps of coal and stone litter the sides. The narrowing roadway slopes down, rises steeply, then slopes down again. Soon the seam of coal shines out at us from both sides; here it is one of the big seams, classed as nine-feet thick in its proper state, but often going higher to twelve, and even fifteen feet. That is too high for comfortable working; the roof is too far away for us to sound it or to safen it with any degree of certainty. Even a small stone falling from that height hits you hard. It is a solid layer of anthracite coal, giving a polished gleam against the greyness of rock and softer stones. A thick black layer that spreads for miles to our left, to our right, and in front of us … Nature's vast store for man from the past. Above us, again, past the layers of stone are other seams … six-feet thick, four feet—three feet—down to eighteen inches, and I wonder how George would feel if he had to squeeze his stout body into that eighteen-inch seam.

Behind us, along the way we have come, the coal has been worked. Steel arches or large timbers support the roof and sides, while some attempt has been made to pack the waste spaces, where coal used to be, with stone rubbish and roof rippings, but we cannot replace so well as nature packed, and now the empty spaces are closing as the pressure from above lowers down on us. The place seems alive, all amove. Steel arches are bending, timber snapping, stones being crushed into dust. The rats scamper around, their movements sounding like the thud of heavy rain on leaves. George has not yet grown indifferent to rats or accepted the ceaseless movement of his underground workshop as part of his life. His face shines with sweat, and he puffs as he hangs his jacket and waistcoat on a nail in a post.

We always have a short stroll around to see what has been done since our last shift. In the stalls, or working places of the colliers, I notice how each absent man has left his smell behind him. The one who ate onions with his snap of food, the other who washes with scented soap, the brilliantine which slicked another's hair, the plug tobacco which another chewed to his content, the minty smell of another's chewing gum, they have all left their scent to remind us while above them all is the smell of human sweat mixed with that of heated coal dust.

Benjy is ready for work and has found his shovel. He leans on it, awaiting our return.

'What is all this they do talk,' he inquires, 'about them anomalies and facilities?' That was not quite his wording, but we knew what he meant. The Porter award was the thing that concerned us the most at that time.[1]

---

[1] Miners' pay during wartime was determined by a National Reference Tribunal chaired by Lord Porter. It made its final Award in January 1944.

'Them's things they put inside engines to make 'em go,' explains Dave, who has also arrived. He is the reserve engine driver and has a habit of beating his arms when speaking, like a cockerel starting to crow. Benjy guesses there is something deliberately wrong about that answer but does not like to ask again. Then Steve arrives with a horse and he gets the usual:

'Where is he?' from Benjy.

'Not come on yet.'

'What time is it?'

'Twenty past eleven,' Steve answers.

'Haven't gone on much,' Benjy complains.

'Not my fault,' Steve assures him, 'and there's a tram of stones in the muck hole ready for you to chuck out.'

'There allus is,' complains Benjy dismally, then as he walks away lifting his short legs over the wooden sleepers we get one of those unexpected twists of his as he starts to sing: 'Do not heed him, gentle maiden.' He has a lovely singing voice and the hollowness of the underground amplifies the sound.

Steve arrives with our first tram and explains the procedure patiently to George; he has done so many times.

'See now? As we shoves this tram round the turn you push this here sprag into the front wheel. Then you screws hard on this front corner. See?'

'That front wheel?'

'Aye. Always when you want a tram to stand on a slope you gotter lock the front wheels. See?'

George puts his lamp on the floor, lifts his belt above his prominent stomach, and sets himself ready. When the tram swings past him he makes a wild lunge, the pointed wooden sprag spins from his hand away from the wheel. George grabs again for it and so misses putting the pressure against the front corner. The tram jumps from the rails and finishes its rapid journey by wedging against a big post with such force that a shower of stones drops around us like the falling of heavy and solid rain. George departs lumberingly to where he judges safety must be.

'Aye. You ain't handling pounds of tea now.'—Steve looks up quizzically to see if the shower has cleared—'and this tram is off road becos' of you, matey. Have a go at floppin' him back on the rails.'

George returns cautiously, faces the tram and grabs the sides in a bear hug. He gasps, grunts, his cheeks become more red than usual, but the tram does not move one inch.

'Like this,' Steve explains. He is about half George's weight, but he knows the way. Digging his heels into the bottom he sets his back squarely against the end of the tram, then with a backward, upward lift he gets that end of the tram a foot in the air and swings it truly into position.

'Easy,' says Steve. 'All you wants is gumption.'

George, well aware that his stock of that commodity is a little short at present, grins sheepishly. It is all so strange to him. Accustomed to be counted a man of substance and to have his opinions heeded, he now realises that he is as a child in this strange world. In his own friendly way he tries to learn from me. He follows me to near the edge of the coal seam and wonders at this vastness of black:

'Lots of coal there.'

'Yes, isn't there. Some of the folk in the big towns would like to have a bit of that—or even what's wasting on the sides.'

'They would that. How do they get it loose?'

I explain, pointing out how the chunks, or slips, run in various directions and how each slip must be loosened before the one inside can slacken. How, when the coal is parted from the roof, posts must be placed as support in its place.

'What's that creaking all the time then?'

'It's the pressure on the coal as the mountain gives down because its support has gone.'

'Oh.' George peers apprehensively upwards gauging how much ground may be between daylight and himself. The whole area quivers under a heavy crash.

'What was that? Explosion, or somebody shooting?'

'No. It's alright. Gas that was, working behind the slips and loosening the coal for the colliers. Don't stand too close. If some falls it will throw back a long way from that big height. It might roll back about twenty-five feet. It hits hard.'

'It do.' George gets well back from the coal. He has not yet learnt to take bumps and cuts as part of everyday life. He had a gash on the back of his hand, and whilst I treated it he stated:

'A lot of folk I know would make a big fuss about a cut like that. Have a week or two off, they would.'

'No doubt,' I agree, 'but you ain't amongst that sort of people now. All you got to watch for is the bump that will put you in shape for being carried out. You'd be pretty hefty, too.'

He shivers, not feeling any pleasure at that prospect. He has not learnt the pitman's philosophy that a miss is as good as a mile. Stones may drop all around you and you may have to sidestep like a boxer in the ring, but as long as they do not hit you too hard there is nothing to fuss about. Even when he is badly injured, the miner's bitterness never seems against the stuff or the work which has crippled him, but to be intensified against the colliery company, or more distantly, the insurance company.

When we get back to our working place I climb on a tram to sound the roof, explaining the danger of the hollow sounding and the degree of safety when the roof rings solid. Below me, George looks upwards, gasping like a fish out of water. At distances of about four feet apart, long posts are placed in position to steady the roof. Their placing resembles a letter 'n' with the bottom spread well out. The upper stick goes close to the roof, supporting it, and acting as a foundation for smaller sticks placed crosswise. The pressure has broken some of these cross-sticks—collars we call them—and one of our jobs tonight is to replace them. Not at all simple, because some bit of them may still be holding, and the roof may come down if it is shaken. The uprights must be fixed solidly so that they will not fall outwards when the centrepiece is taken away, and the new collar must be gauged and measured in relation to the distance from the roof and the bends in the nine-foot post that we intend using. We have an extending stick with which we measure, and placing it in position we calculate how the bigger timber will fit when it is lifted up. Being so heavy we do not want waste liftings.

George cannot help in this assessing as he has not yet learnt the knack, and even when the lifting comes he must be watched because he cannot be trusted to lift with the exact steadiness at the right time. When six men are lifting their utmost it is necessary that they lift together, or perhaps an unbearable burden will come on part of them. In a position where inches are of immense importance, the slightest bend in a long post can make a big difference in the space it needs. That is why we study the shape of each post so minutely.

We cannot travel many yards in any direction without having to push our way past the brattice cloth which is placed to turn the air current. Heavy, clinging stuff it is, reeking of tar, and our skin burns if it rubs against this cloth. The airways spread about like the veins in a man's body, and we can find our way by throwing up a handful of dust and watching which way it blows. Back on the roadways we follow the narrow tramlines.

Later we make room for more timber. This needs patience to saw old wood away inch by inch, sometimes with the saw high above our heads. Then we have to widen the sides, slowly tapping and breaking the stones, and packing with wood chips or smaller stones to prevent a slide of rock or other rubbish which would take us hours to clear. Our job is to keep the place standing, not pull it in, and any rash working might mean a big fall.

Benjy creeps around occasionally, greatly concerned lest we should draw a lot of rubbish and nearly always bringing the question of 'What time is it?' His visits must be swift and furtive, for there are very few minutes during the shift when the fireman is not near our elbow. The fireman is either terribly conscientious about the work or very much afraid of losing his job. I have never yet been sure which it is. Yet he is one of the best officials I have known, a really decent man. He would be horrified at the idea of forcing men's pace when on a dangerous job, yet that is the effect of his constant visits. He has less than a dozen men in his district, and so most of the interest is centred on our particular job. He may be lonely, or he may fear falling off to sleep; and so his antidote is to remain close to us, having a word now and then. One night the overman sat with him for some hours while we worked a few yards away. Two better-paid men watching two others sweating—so very pleasant.

'Like a blinkin' bo-peep about here, he is,' Steve says.

'Rotten thing to watch men at work,' George complains.

'Some folks seem fond of it on the street outside,' I suggest, looking at George.

'Oh, yes.' George is confused for a minute, possibly his memories are guilty, 'but it's so different outside.'

In a way it is. The trouble underground is that danger is always lurking and the presence of an official has the effect of making some men run risks they should not. The other argument is that the official is there to help watch for danger and give warning. I once saw a man smashed by a falling stone before the watching official had time to open his mouth for a warning. Had no official been there that man would have trusted his own experience and not gone under the stone. He might have still been alive ... but he had a family. So I do not take kindly to being watched, and by wide sledge swinging and sharp backward jumps try to show the official that we want more room. Yet when I jump squarely on his toes he apologises to me for being in my way. What can a tender-hearted man do in a case like that?

When the earth is shaking itself after its nightly slumber we have our meal. Steve brings the horse along and he blows his chaff as we eat. He gets also a bit of spare bread and margarine, but Steve collects the scraps for a cat in the stables underground.

'Poor devil,' he says, 'I haven't seen nobody but me giving her a bit of grub. Got some kittens, too.'

We take our full twenty minutes, watching the last seconds. Benjy has a nail sticking up in his boot and arranges the lamps around whilst he pushes a sledge inside his boot and hammers the outside with a hatchet head. 'Any more cobbling to be done?' he inquires.

Then he becomes annoyed as he tells us of a woman of ninety-three who is getting good wages at a Government factory. 'Two p'licemen helping her to work every day, aye muneferni.'[2]

'She can't be ninety-three,' I argue.

'Over eighty, anyhow,' he insists. Later he comes down to knowing that 'she is a good age, whatever.'

Soon after came a crash which made us all jump like puppets under automatic control. When we had found the point from which it sounded a large piece of side had fallen against an empty tram and had buckled that tram badly.

---

[2] A Welsh exclamation: *myn uffern i*, equivalent to 'for heaven's sake!'.

That meant getting the fall away cautiously from the side of the damaged tram and starting the tram on its way to daylight and the smith's shop. Fortunately the wheels were still turnable.

George looked long at that bended sheet of iron which formed the side of the tram.

'If it would do that to iron what would it do to a man's body?' he asked.

'Forget it, George,' I advised him, 'the marvel is not that so many are hit, but that so many escape.'

Soon came the time when the chips were cleaned up and the roads tidy; then the tools went back on the locking bar. Steve started to drop the harness of his horse and Benjy came round the corner with: 'Where is he? What time is it?'

'Gone to make his report,' said Steve, 'and it's time to go home.' So we went gladly.

We walked into the dust from other miners' steps as we travelled along the two-mile journey. Outside the day was dawning fine; the birds welcomed us from every one of the many trees on the lower mountain. Those trees were still bare, but there was a promise of budding and a feel of spring. The outer world seems sweet to us when we come out.

The other shift came up to meet us, white-faced in contrast to the coal dust we wore, and near the canteen the laggards of the day shift talked with the first of the black crowd that hurried down. Another shift was complete.

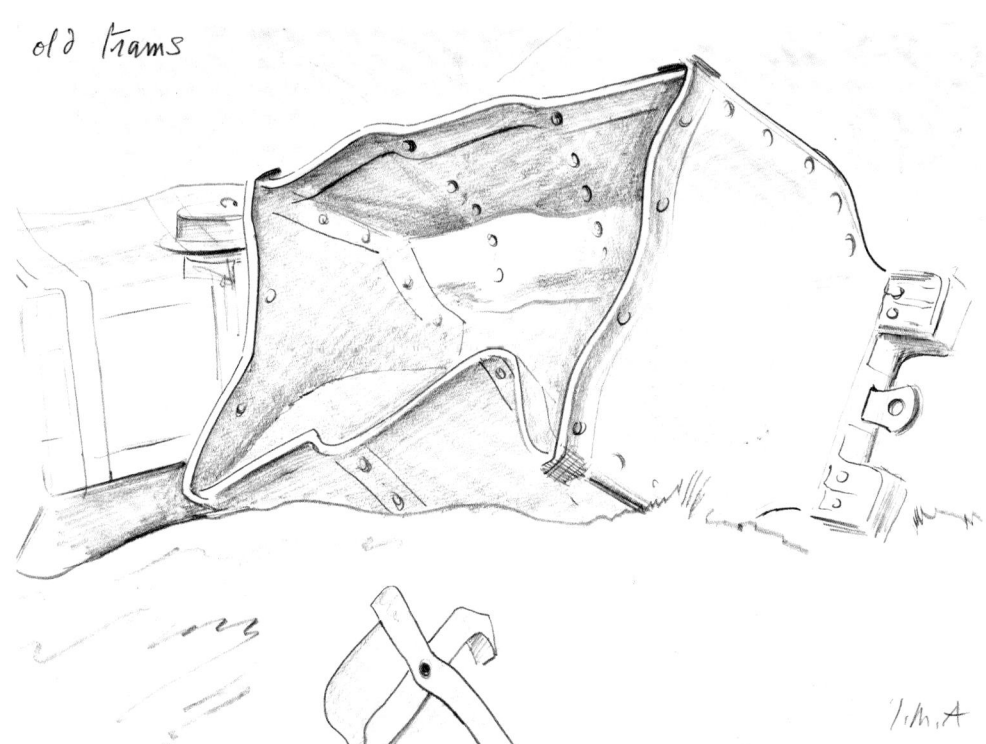

Old Trams, c. 1943–5, pencil, 22 × 31

This is a sketch for a lithograph of broken trams of riveted wrought-iron plates used as an illustration in *Miners Day*.

CHAPTER TWO

Jerry had always been useful about the house; he had a hatred of idle hours. You noted that in the neatness of his garden and in the fullness of his homemade glasshouse. His fowls had an attractive cot and a good hanging of the greenstuff for their picking. His skill as a cementer showed along the paths and about the pigscot. Yet of recent years there had been many idle days for him: days when he was like a cork tossed by the waves of Health Insurance of Workmen's Compensation or Parish Relief. He had no solace or welcome from either. Each threw his appeals along to the next, and the skin stretched even tighter on his cheekbones and the brightness faded from his eyes whilst Officialdom quibbled as to whether it was a weak chest or bronchitis, an overload of coal dust or the scheming of a malingerer.

'Sending the fool farther, they are,' he told me, a gasp pushing each word. If a fool is a decent man, willing to work hard for a fair living, then Jerry was one. The period came when it was an effort to speak or move. He would not lie propped up in bed, or even sit his last days away in a chair, so he struggled to put a porch over the front door and put a rough seat of boards inside it. There he hoped to sit through the spring, watching the garden become fruitful, hearing the birds in the plantation and seeing the brook toppling over the rocks just in front of his door.

As always, he had builded well. The sticks were slotted as he had placed them in the mine, and the boards were firm. I sat on them whilst waiting for the funeral to start. His slicing was still fresh on the uprights and his dog wandered about looking for his master. Part of Jerry had been taken for a final examination and we were going to carry the remainder on his last journey.

It was rather early in the afternoon, so we were not more than thirty at the start. Workman's buses going to the distant areas slackened their speed and went silently by. The occupants bared their heads as a last tribute to the man they had known. From every side road men hurried with soap-shined faces, and hair they had not had time to dry. Soon we were a long procession tramping grimly on.

It was a little church, very old, very lovely—with the graves of the landowning family set in a place apart. Most of the mourners entered with trepidation and sat in stiff unease through the service. What thoughts must have disturbed the minds of these men, who were all aware that their turn would surely come to surrender to the dust menace! The four young sons, one in khaki, were sitting near the coffin. They had seen their uncle die from the disease, now their father. One of them was already badly affected, and another, little more than a boy, was beginning to suffer. What thoughts passed through their minds? Possibly it is kinder to hope they did not think at all.

'Dust to dust,' the words and the token fell on a man whose inside was hardening into a cemented mass before he was half through life's span. The monotonous service, the aloofness of the church, is that all Jerry was worth? With such great need and such opportunities, why does the Church seem so ineffectual? That evening was mellow and bubbles of green showed along the hedges. A waterfall sparkled and slashed near the graveyard, the mountain was covering itself with a green dress and a cluster of birds in the fir trees sang a gentle farewell to the day and the man. Yet the earthly representatives and symbols of such a Creator droned their duties away and appeared as feeble as man can well be.

Using a short cut, away I went up a little slope and paused at a stile which overlooks the churchyard. I could not cross because an intimate friend of mine was leaning against it and he wanted me to linger; besides, I knew he could not now move quickly. Dan has one of the most expressive faces I have seen. Yes, it has become haggard and that blue tinge shows on his cheeks, but there is always that inspired look about him which convinces you he is thinking of something which seems wonderful. He has a rich, tuneful voice and his eyes have the sparkle of an enthusiast.

'So Jerry's gone,' he says.

'Yes.' I agree, inadequately.

'One more,' he continues. 'I wonder is that the end?'

'They tell us not.' I feel uncomfortable in this discussion. I know, and Dan knows, that we have been watching a ceremony which will evolve around him within a very few weeks.

'They tell us,' his voice is disdainful, 'they tell us all sorts of things. It's a living for them and a soft one, too.'

'Coming home,' I suggest.

'Not yet,' he explains. 'I'll take my time and I've got nothing to do.'

I remember Dan coming up the slope from work on his last shift. That sharp slope is a sort of barometer in our lives. The younger men hurry up it swiftly, leaving their dust floating behind. The more mature have steadied and climb with a thudding dig as they put each foot firmly down before the other moves; they have learned the need to save their energy as the crest seems to become more difficult to gain. Then you notice others coming up more slowly, pausing on the way, and can notice the panting lifts of their breathing. Soon comes the time when they doubt their ability to reach the top, and they know this will be the last time their chest will stand the strain of climbing.

I had watched it coming with Dan. He was getting weaker and more easily exhausted, yet he knew no alternative but dragging himself to work. Then that time when we had to help him up and he came no more. They fear going to their doctor because he would tell them to cease work, and they have no other way of living. For an active man it is a taste of hell to be told his use on this earth is finished and he can no longer provide for his family. If there was some very light occupation we could put them at it would ease their minds and they would not have the chance to brood, but there are no light jobs in mining that a disabled man can do. I know one colliery where sixty men are waiting for light employment—unfit men, I mean.

The full compensation is only a starvation payment, but to get that amount a man must be certified as completely disabled. A Medical Board gives that decision, but between the time of a doctor warning the man not to go underground again and the sitting of that Board anything up to a period of five months may pass. Until he is certified he gets no compensation; when he ceases work his wages stop. What can he live on during those five months? Usually it must be parish relief. If, in the feeble hope that he may obtain light work, he goes to the Labour Exchange, it would be necessary for him to get his cards from the colliery office, and that could be interpreted as breaking his contract with the company with the possibility of absolving them from liability.

If the Board, on the X-ray evidence, find that he has not swallowed a sufficient amount of dust, he may have to go back underground to complete the process and be examined again in six months, or in more severe cases he is given partial compensation and advised to get other work—which must mean taking his enfeebled body away to other work and amongst strange conditions.

That night in work I was told that another investigator, I understood it was a Means test man, had arrived at Jerry's home an hour after the funeral. Benjy had a shock that night. It was docket night, when the payslip is given out and most of us wear a glum expression. Benjy had his and hurried away, complaining of something that had happened. About fifty yards along he decided he would see how much pay was coming; looking at the paper in his hand he made one of the queerest noises I have ever heard. He turned so swiftly as almost to collide with me and said, 'Look what the sods have done to me.'

Before I could look he had rushed to the fireman, exclaiming: 'Hey. This ain't no good to me.' His paper was quite blank, without a figure or a name. We all studied the white sheet and not one guessed the solution. Benjy was a dancing bundle of profanity. To calm him the fireman promised to show it to the overman, and was folding it up when he noticed figures on the other side. We had all been looking at the back. Benjy went off pacified, but still complaining that, 'it don't make much difference which side you looks at, whatever.'

The Porter award had been our chief topic of late. We had asked for a six-pound minimum wage, and after an unusually quick inquiry we had been given five pounds. The papers which had coaxed us into thinking that our work was of value and our quest for an increase was just, suddenly switched over to telling us that the five pounds was fair, and that peace was assured by it in the industry for a long time. As it meant that a large proportion would get no increase at all it is difficult to see where they found that comfort. Our local leaders did their best to soothe us by claiming that no one would be worse off—a rather queer claim, for we do not usually ask for wage increases with the intention of being worse off. The day that this five pounds award was announced to us the wireless gave the Board of Trade figures, which stated that the average wage for an adult manual worker throughout the country was six pounds and one shilling a week. Feeling was bitterly smouldering in our mines and homes that weekend. During the first shift of the following week, news had come through that all our customs and privileges were to be cut and merged into this one wage—although we had previously been told that this would not be done. It did not need much experience to tell that trouble was about. Even George could sense that.

We were working at a place where the steel arches had been buckled up and were bending dangerously. It is first necessary to unbolt the fastening plates, a risky job, as they fly back when released and may travel through the air for fifteen or twenty yards; quite liable to harm harder things than heads in that buzzing flight of iron. The steel arches are sometimes forced a yard or more into the earth and are bent at all sorts of fantastic shapes. Not only is it difficult to get them free, but there is also the problem of getting them in trams and out to the surface. Space is

Old Rails, c. 1943–5, conté, 26 × 31

Despite the title, these are not rails but twisted steel roof supports. A similar image appeared in Miners Day. Much of Bert's work repairing roofs underground involved having to remove damaged supports and replace them.

limited and their bends are not. They are nasty things to handle as they slip and crush any unwary fingers; besides, the surfaces get roughened and bits of rusty iron pierce our hands and bodies.

I feel they are better support than timber, for they rarely snap and do still hold some weight even when bent. We had dragged out a couple of arches and were preparing for the last. The timber laggings were hanging slightly downward and I was watching alertly in case they might slide and the loose stones above follow them down to where we were working. I put George on the upper and safer side because I wanted room to jump back, and George, despite all warnings, has not yet learned to give his mate a clear backward passage. Usually, when I jump back I crash into George, who has just then decided to come behind and have a look at what is going on.

It did happen suddenly, I know, but I was half prepared and was leaping back at the first crack. The road was rough and some empty trams were rather close. For a few seconds it seemed the world had closed in on me, and I instinctively rolled down alongside the trams for their shelter. Roof and laggings came down with a crash, forming a huge mound under which was my lamp. As something, stone or timber, had caught me nicely in the solar plexus and knocked every pant out of me, I lay awhile in the dark, helpless and speechless.

Then I heard a horrible sound, not easy to describe. It seemed not human, a sort of animal moan mixed with a human scream. Some seconds passed before I realised it must be George, and that he must be hurt. More time passed before I could stand up and croak the message that I would be there in a minute. Then when I struggled forward and could see over the fall, I saw that George was not hurt but was running back and forth in his panic. He became more normal when he realised I was not badly hurt.

Poor old George! The crash, the silence from me and the disappearance of my light had convinced him I was buried. His first experience of that type of happening, he had no idea what to do. Later I gave him some tips from my experience. Not to attempt to help yourself in case of more falls, and because one man is usually impotent against large stones. Get the message to others quickly, and they will be there very quickly, never fear about that.

About six tons of stone and a big stack of timber had fallen. We got that lot clear by finishing time. My lamp was not damaged, as stones had arched across it. George went home thinking deeply, with another new experience in his mind; an experience that has become one of the ordinary happenings of life to most of us.

That morning we heard that many South Wales collieries had ceased work and the others in our valley were idle. A meeting of our men was arranged between the day and afternoon shifts, and I was there in the crowded hall. Grimy day men came in off shift, afternoon men ready for their shift, and a considerable number of night men, heavy-eyed because of their broken sleep. The large hall was abuzz with argument. The meeting was conducted on strict lines. The colliery chairman and secretary each spoke, giving a fair report of things as they understood them. Everyone then had a chance for his say, and the chance to put questions was taken liberally. The argument that a strike would let our soldiers down was countered by men who had brothers and sons in the forces who, so they claimed, had urged them to fight and maintain their customs and privileges. They argued that they must retain something for those absent ones to come back to, whilst the suggestion that we should wait for further negotiations was swamped by the reply that we had already waited a long while and that we had disputes still unsettled after twelve months' negotiations—and that is still very true.

Feeling ebbed and flowed whilst outside the sun shone and the buses waited to see what the men would do. The scales were loaded against continuing work, because all around us collieries were idle and we felt their fight must be ours. Yet it was only a small majority that decided we would not work. Once taken, however, the decision was accepted by all without the least quibble. They streamed out, and homewards. The miners have an extraordinary loyalty to one another in matters like this. Nothing would have induced any man to work in defiance of that decision. Possibly our action was wrong in view of circumstances, but it was but the climax to a series of happenings which should have been avoided. It was the outlet of a seething disgust for the continual delays and evasions, intensified by the feeling that men who knew nothing of our work or ideas were making decisions that would affect our lives and our families.

The night shift came along from their homes. We felt they had a right to their opinion, even if that voting could not alter the result. They sat in rows in the canteen, with a cup of tea or some other drink before them, heard the report and approved the decision. Then, by bus and train, they returned homewards to wait until the call back to work came.

Things became quiet. With the colliery idle the local folk could do their washing and even see the green leaves emerge from the grey dust mass near the colliery screens. The air was sweet for a few days. The only excitement happened when a man who fancied himself as a wit called our local coal haulier a blackleg because he was taking out coal. That was the unforgiveable sin, as the workmen's committee had given permission for the coal hauling to go on. The driver, an elderly man, reached for his shovel and brought its width down forcibly on the head of the accuser. He paused long enough to be sure that the blow was enough to keep his traducer on the ground for awhile, then resumed his coal delivery. Everyone, except the man who had been under the shovel, thought that honour was well satisfied.

Friday came, bright and fresh, with the wet coal dust lying still in the gutters. I saw all my mates in their best clothes, and they bore comparison with any other workers. We had our first payment under the new award. I gained three shillings a week and my son Peter got four shillings advance. George now gets exactly the same money as myself.

Another instance of five hauliers working together. One of them is given a turn a week to see that trams are shared properly. They take that in turns, so each one gets an extra shift every five weeks. Without that shift they are under the award, with it they are over it. To add further complication three of them receive coal, bringing in that problem of one-and-sixpence a week, whilst the other two do not. How are you going to square this little lot up?

Then we have the rope riders doing extra work, such as shackling, loading timber, bringing out late journeys, loading steel arches and other jobs on which it used to be the custom to employ extra men. For that work they were given an extra shift a week and the company avoided employing special men for the jobs. This award merged that payment into the total to make up the minimum and the riders felt and said that, 'Lord Porter had best come and do the shackling himself.'

Then there is Jim Lamey, who was caught with a fall and was in bed for more than a year. He now drags part of a fine body behind him and does work at the labouring rate. Apart from his ordinary job he used to make a painful journey through the workings to drive a small engine when it was needed. It prevented the cost of keeping a driver waiting there for hours and Jim was given half a turn a week to help him. He also got a small sum as partial compensation ('compey' to Jim). The Porter award has taken both those from him and he now asks: 'Hey, you. What about my half turn and my compey? Won't get me dragging this old leg about the place if I don't get something for it, you know.'

So the problems came in from all sides. The secretary, with the rest of us, spent all our days trying to ease them, and were repeatedly told by some of our mates that we were doing all right out of the strike. We were entitled to a payment of one shilling for each committee meeting. Frequently that meeting took a slice out of the day, and about then, as the funds were low, we decided to forego even that shilling.

That one-and-sixpence deduction for coal was a bitter pill; we taste it still, and have not forgotten what it means. Cheap coal has been our privilege: about our only one. Most trades get their privileges in the way of cheap clothes or travel or some allowances; cheap coal was ours. My ton load cost me about fifteen shillings at the door, and I get an average of eight loads a year. Lord Porter has assessed an added value of three pounds eighteen shillings a year on that coal—running near to ten shillings a ton. In the case of a man living in apartments and allowed only half that coal, its value to him would mount by nearly a pound a ton.

If that extra eighteen pence was put on the pay docket and then deducted, income tax would be taken from it, and so a man who earned four pounds eighteen and six before the award would still receive the same amount, but would have to pay an extra pound income tax, rebutting the claim that no one will be worse off by the award of wages.

Over that weekend our mass meeting decided that we would return to work. The old situation had again arisen. Month after month passes whilst a dispute is rolled back and forth, then the men strike in weary desperation. Instantly they are informed that nothing can be done while they are idle. If they return to work negotiations will be hurried. In this case as in many others, the hurry should have taken place before the stoppage. We felt we had made our protest and there was a relief in returning to work.

I think Lord Porter made a sincere attempt to solve the problem. Miners have asked for a flat rate and a better deal for the lower-paid worker for many years. Porter tried to give this, and he did help the underdog in the mines; his award would also wipe out many of those puzzles on the pay docket. The mistake was in not recognising that there is skill in mining. You cannot expect a miner to give the experience of thirty years for nothing, any more than you can a doctor. If he has to take responsibility he wants to be paid for it. If he has to buy about twenty pounds' worth of tools and maintain them in good condition he needs a bit more pay than the man who has only to purchase a shovel. The Porter award had some good points, whilst in other ways it went off the rails.

In their past clamour for a flat rate I think that the miners misstated their case. What they sought was a wage which would make piecework unnecessary with all its added risks and unfairness; and they wanted to do away with the system of allowances which offers such chances of reward for tale-tellers and creepers.

Benjy is at it again. He has lost a couple of days and comes towards me waving his docket. 'One penny rise I've got,' he shouts, and it is true. 'One penny rise that old Porter gave me,' he complains. 'Blinkin' good job he wasn't a stationmaster.'

Blaencwm, c. 1943–5, watercolour, 24 × 20

This watercolour shows the incline that carried spoil up the hill from the screens of Glenrhondda Colliery at Blaencwm. The end of the derelict Lower Terrace is almost clipped by the tramway, which went up past the tram sidings at Hendrewen shaft to a tip on the site of disused levels under Craig Selsig.

CHAPTER THREE

'Another blinkin' miner's weekend' was the emphatic grouse from some of the men on a Saturday morning as we came out to see the rain swishing over the loaded coal trams and damping the small coal underfoot into a paste, which hung on my boots as I squelched my way homewards. I pondered over that complaint. It sounded as if they considered this dampness would be a concern of theirs alone; that it would not hinder the pleasure of anyone else.

I have noticed that mentality a great deal amongst miners. Possibly it is due to the seclusion of the mining valleys and to their limitation to one industry. Mining is the only thing that matters and the sole industry to be considered. The mining valley was their world, and as they thought and felt so should the rest of the people. The idea is losing its grip during the present period, but it has caused a deal of misunderstanding. We need to realise more fully that when the bell tolls for the cotton worker or the farm labourer, so the distant echo of its clang will quiver into the lives and earnings of the miner.

I recall that Saturday evening because I was going homeward through the rain, and a ripping wind, when I passed a small wooden chapel. I have my own routine on Saturday afternoons. About five o'clock I walk across to the Welfare Library, which is about a mile away. There I go through all the dailies, balancing the views of each against the others, then the weeklies, with a special long perusal of *John o' London*'s.[3] Having satisfied myself that no new printed word is there available that I have not scanned, I go into the library, argue about books and the affairs of the country with the librarian, and then select two books; one for myself and one that I think my wife might like— always making sure that I will enjoy it in any case.

Satisfied then with a packet of cigarettes, I start for home, leaving the pubs and the pictures free for that night to those who want them. My wife will be prepared, with the oil lamp lighted and a fire of wood and coal blazing well. Then after tea, two chairs alongside the fire, and silence whilst the radio gives us Saturday Night Theatre; then the books until late, when the boy arrives home. That is our Saturday and we would not change it.

I was looking forward to that fire as I splashed along. There were no half measures about that squelching. It had been dry for some weeks and over that area the pit dust falls continually like a fine dry rain. It was surely a couple of inches thick, and then came the real rain to hold it close to earth and soak it well. You could see every footstep that

---

[3] A widely read literary weekly from 1919 to 1954.

Dinas, At the Backs of Houses on the Riverbank, 1944, pencil, 21 × 27

This drawing was published in 1946 alongside the essay 'Coal: The National Plague Spot'. Terraced houses are squeezed above the railway and the river. One row (probably the since-demolished George's Row, Trealaw) has long gardens, and on the near bank a strip of waste is fenced with old bedsteads to make a space for laundry poles and a home-made shed. A woman carries her baby in a shawl.

had converged on that little chapel. Huge number tens and dainty high heels, all had left their deep imprint, not on the sands of time, but on the dust of that mining road.

There were other marks, and passers too, as I sheltered alongside an old canal wharf from a shower. Two women, carrying babies in shawls and a bucket full of something resembling coal, dragged themselves through the wet dusk whilst two men and a young boy strained to push a wooden cart with a tyreless wheel through the clinging slush. They had the mark of men who, though not even middle-aged, were past doing mining work, and they had been digging into an old tip for some fuel that would keep a smouldering fire over the weekend. Above them the great mountains were thick-lined with coal, and a hundred yards away three railway engines belched smoke and cascaded sparks as they started the wheels turning of a line of about fifty loaded coal wagons—say five hundred tons—on their way to the outside world.

Then from that chapel came the singing. Nearly always there is a drag about that sort of singing; the organ is just half a beat in front of the most confident voices and the others tail behind like the lame in a procession, but what I heard that wet evening was virile with an exaltation which sent a thrill through my being. It was the expression of a happy people, and it made me feel they believed in what they sang. Never before has hymn-singing affected me in that way.

Whilst I sheltered alongside the wall a companion sidled up to me. He was about six years old and was eating an orange—an unusual occupation in these parts and at that time. After a consideration in which he decided I could be trusted, he informed me:

'Lot of them Yankses been goin' by.'

I agreed, having both seen and heard the convoy. My companion kept digging his teeth into that orange, blowing and gasping as he did so. The yellow from the fruit stuck closely to the dirt on his face, but he seemed intent on finishing the eating before he went home.

'I did shout,' he informed me between fruity gasps. '"Give us some gum, chum," and they threwed me this n'orange.' Away in the distance a lorry backfired and the small boy was entranced.

'Hear him pressing his button?' he questioned. Obviously he was a picture-goer and an aeroplane student. I waited to see him trotting homeward to some grey houses, transferring the juicy evidence from his mouth to the back of his hand. The exuberant singing went on as I squelched homewards; an island of joyous sound in a dreary, almost silent night.

Sunday in the mining areas can scarcely be anything but dull. Even the Welfare Halls are closed, and in Wales, of course, so are the public houses. Save for a stroll along the road in fine weather, or a hill climb for the more energetic, there is nothing to do or see. The only doors opened lead into the chapels or churches, and even with this compulsion the attendances are dwindling every year. In many cases it has come to be thought of as a social occasion where new clothes can be aired and old friends met. Habits hold tightly, and when the journey has been made for some years it is often easier to go than to stay away.

There is a feeling that one is part of a community and of some value in the sight of others. Once a break comes in the attendance it is usually the final parting. Another angle is that of the small shopkeepers, who go to court the favour of influential customers, and sometimes a certain type of miner attends regularly because he knows the higher mine officials will be there and will notice him. This fact, that the mine officials will be prominent, each in his own special branch of religion, at once presents a reason why the most politically-minded or class-conscious of the workers will not attend, and no doubt places a handicap on religion which is not always justified.

One Monday night we had arrived underground at our work, and were studying the best way to do it when the fireman came hurriedly with fresh orders. One of the stall roads was crushing badly, and this action was endangering the main airway. These airways are as essential to a mine as a throat to a man. Once they are completely blocked and the flow of fresh air cannot be drawn through, then the men must be rushed outside. Within a very few minutes gas would be accumulating and it would be dangerous for anyone to approach. Gas is probably always present in this type of mine, but the continual flow of fresh air dilutes it and takes it along. Once that dilution ceases things become very awkward indeed.

We got our tools handy, then fetched two large posts. There is no joy in the effort of dragging heavy posts through the twistings of old mine workings. Arrived at the job, we slid up the measuring gauge to the necessary height and hurried back to cut a bit off the first post. While we were cutting it that part of the roof under which we had measured crashed down. That post was useless, as there was now about twenty feet of height where it had been nearly nine feet before. The airway was blocked, so when the falling had ceased we crawled continuously forward and rolled stones away from the passage. We climbed over a small mountain of stones and slid down the other side to the airway. Above us things were looking ugly, very ugly. Some stones were hanging half fallen, waiting just one more slight jerk to complete their smashing fall. Others wedged one against the other in their dropping, and still waited above us, many of them a yard thick and more than six feet long, as if undecided whether to stay up or finish their drop. In such times it is all a matter of luck; you cannot get a chance to put any supports. If your luck is good the stones hang while you scramble through, but if your luck is out then the stones may drop as you crawl underneath.

Sometimes you get a warning when little stones drip before the larger ones start. That happened, and I made a leaping tumbling rush over the piled stones back to George and our other mate. Only one of us went forward at a time, so that the others would be back clear to help or to fetch help, and to give the escaper the most room possible. Before we got to the airway we had to scale a nine-foot wall of rock stones, and that was one of the problems. My lamp weighed nearly thirteen pounds, and I had to hang it up so that both hands would be free. In the sudden frights I had to jump straight backwards over that nine feet into blackness and on to hard ground, or perhaps timber.

Our extra mate was an elderly repairer, who could

Blaencwm Level, c. 1943, conté, 28 × 23

One of Isabel's three known underground drawings shows a roadway with a high rock roof supported by large timbers. Loose stone is packed to the sides of the tram rails.

be relied on to do the right thing at the right time, but was past his climbing and leaping days. George, very nervous, but desperately anxious to help, would surely do exactly the opposite of what we expected him to do. That way it did happen when some small stones dribbled right on my back and I retreated at a jump. There was not time for a cautious clamber, so I jumped backwards and took the chance of an injury that way. In the same moment George jumped forward to help me down. It was a mild example of an irresistible force meeting an immovable object. Some minutes later, when we had both recovered our breath and our tempers, I began to ponder whether it would have been less painful to let the stones hit me.

Just after that my lamp was buried, but we got it out safely. Later, twice, the tools were completely buried, but we got out undamaged. A huge flat slab dropped and we saw it as a foundation for some supports. After many furtive dives forward and blind backward leaps we managed to get four large posts in position; sloping at such an angle that the lowering roof would meet their tops squarely. There was a gap of eight inches between their tops and the nearest part of the crushing roof, but we knew that space would soon be lessened, so we left the posts shored up and waited. Less than half an hour after, the roof had lowered that eight inches, the posts were holding solid, and the weight was squeezing turpentine from them.

Seven times during that shift the airway was blocked; each time we got it clear in time to prevent any stoppage at work. Thrice as often we had to make that perilous leap over the high wall of piled-up rocks. My arms and shins were shorn of pieces where I had misjudged the landings. One time, when the whole underground area seemed to be amove, we retreated well back. As it did not ease we decided to eat our snack of food to give things a chance to quieten. A rock-and-timber-cracking symphony was going along nicely even fifty yards away from the worst place, and we looked around and above very carefully before we sat on some stones to eat; crouching against the side to present as small a target as possible. Food finished, with the cracklings all around us, I closed my tin and placed it near my elbow. A few seconds later a stone dropped square on it, crushing the tin into the ground. That means another new food box, and their price is three times what it was before the War.

'No blessed sense in men having to work in a place like this,' George complained later, in the intervals of running in to work and dodging back from falls. The elderly repairer, well seasoned in such happenings, looks on with the stoicism of a farmer who sees the rain prevent him working; but the rain in this instance was hard, made of stone. It was about five in the morning when the commotion slackened, and true to form it had been at its most violent from three to four o'clock when the earth is supposed to be waking for another day.

When the falling slackened, we went in again to make a final clearance; one watching above whilst the other cleared the blocking stones, and we gasped with relief at the coolness of the inrushing air. We dare not use a sledge lest the jarring should start a fresh movement of roof, and we spoke in hushed voices as it seemed the vibration of a word might bring more stones down. Gently, almost crooning to the stones to behave, we cleared the narrow roadway and covered it with timber so that it should be secure until a better cover was arranged. A quarter of an hour before finishing time the fireman came along, and we gave him some details of our trouble. The airway was clear, and that was all that concerned him. He had no need to be told of all our difficulties, because he had enough experience to sense them; they are part of mining life. As we had some minutes to spare he suggested a couple of little jobs which needed doing, but there was no conviction in his orders. We felt we had done enough, anyway, so those jobs were left for another shift.

George was indignant. 'Been going through hell all night,' he complained, 'and now they sets you another job for the last couple of minutes. Folks get medals for a lot less than we have done.'

'We won't get any medals,' I assure him, 'we're lucky if we get our pay all right. Don't worry about extra work. He only said it to show his authority. It would be when he tried to make us do it that the band would play.' George was still unappeased, and stamped about with the sweat shining on his face and his shirt wet and sticking to his back.

It made me recall one similar night in another district, and with another fireman. The first shift after some holidays we had a terrific night, and none of us were in real shape for it. The fireman had begged us to make a supreme effort to clear a main roadway so that coal could go back on the

MINER'S DAY 71

morrow. We worked like madmen under a bad roof of every minute of the shift, and finished, making a good job of the repairs, a quarter of an hour before the end of the shift. The fireman arrived, saw that we had done even more than he had begged us to do, and went away without saying much. A couple of minutes later the rider arrived, grinning, bringing the message the official had been afraid to give himself. There was an empty tram farther back, and we were to pull a section of roof down to fill that tram. That was to be our encouragement for slogging.

Nobly assisted by the rider we sat down, five all told, and explained to one another what the fireman could do with the tram, what we thought of him and his relations, and of his past. We scorched the atmosphere with our language and our suggestions, and the tram remained empty. A few minutes later, when it was time to go home, I doubled into an airway right by us in time to see the fireman slinking away. He had crept up to hear our reactions to the delegated orders and had heard them with full embroidery. Working life was very peaceful for us in that district for a long time after that.

House coal, that cheap monthly allowance which is about the only privilege that a miner gets, has been causing us a lot of trouble lately. The Company have not been delivering it regularly, and the men are annoyed, as it is part of their wages. As usual the 'pass it on' method is employed. The committee see the coal clerk, who says he has his orders from someone higher. That one is seen, and passes it up to another, who passes it up again, and finally, when it can go no higher, it is pushed back to the coal clerk—and trade union representatives always have to be cautious not to put any definite blame on a workman if they can avoid it. Some weeks ago we had a short strike over this cheap coal question. It concerned the disabled men who had been having a supply of cheap coal. There was a suggestion that this allowance would be stopped and the Federation officials reported this to the men. They went home in protest. After some lost days we resumed work, and now the same question was coming up again. It seemed well set for another stoppage, but we were at last informed that all disabled workers would receive their cheap coal until the Medical Board should decide their case, and the whole arrangement was to be settled by the Conciliation Board. Of course, when a partially disabled worker accepts any other work, his cheap coal stops. Also when a man has been killed underground his widow, under most companies, stops getting any cheap coal. I know of hard fights taking place to get the coal for a dead miner—coal that should have been delivered to him before he was killed.

I notice a side issue on this question when a man who had been a colliery official for over thirty years, and had finished through ill-health, had a load of coal taken to his home and tipped. Then someone found he was not entitled to cheap coal. He had been working up to a short time before at the colliery, but a cart was sent and the coal loaded again. The miners treated this as a joke, but if it had happened to a workman there would probably have been a riot. The men sense the gap which opens between themselves and the officials. They have no interest in the way colliery companies treat their old officials. 'They've been handy for them in the past, let them look after 'em now.'

Trouble threatened also from the hauliers and riders. 'When are we going to get that half-crown?' That half-crown is in a claim that has been debated for quite twelve months. It is a weekly matter and concerned with the Greene award.[4] After the first preliminaries were over without any settlement, it was sent away to headquarters many months ago, and yet no decision has come. Our leaders show us proof that they have been continually trying to force a settlement, but it is always postponed or evaded. I once thought these days of blockage and deliberate delays were over, but it does appear that things go on in the same way no matter what regulation is passed. The result is that the workmen put most of the blame on their own leaders, and state definitely that they must be sleeping about such things. About once every fortnight we have a lively meeting about this half-crown, and our secretary clears himself by showing all the correspondence. There appears to be a definite strategy of delay: and the owners exploit it to the full, knowing that time is always on their side.

One reason is that both sides have become so closely amalgamated for attack or defence. A decision in one colliery may, in the end, affect all the collieries in that area, perhaps in the country. As we have, or did have, a short while ago, in this country, two thousand one hundred and twenty-five mines under the control of nine hundred undertakings, I can easily understand how those who watch, very keenly, the coal owners' interests are prepared to fight hard before they grant any concession or even any right which can be disputed. This question of half-a-crown, if given weekly, might start similar claims at some other collieries, perhaps at all. In a similar manner any alteration in working methods at a particular pit must be watched in case it should be introduced in, and be detrimental to, the workers in any other area. It is obvious that there is no exact precedent for the payment of this half-crown, or our representatives would have found it and argued their case along that line. Yet it is a just claim and has partly been admitted as such. Between it all the months pass until the men, seeing no reason for delay, become weary and feel they will never get a decision. Usually, when tempers rise and work stops, a settlement of the point in dispute occurs within a few hours. No wonder the men are dubious about disputes machinery.

A badly damaged lamp had been given in and a prosecution was threatened. The workman said it was an accident and there was no definite proof otherwise. The company argued neglect. The fine was heavy: five pounds and costs, so I believe. Then the whole colliery was seething. 'What's the use of our Union if we lets them get away with that sort of thing?' The truth was that we could do nothing, as the workman had no proof that he had reported the damage and for not reporting

---

[4] Lord Greene, Master of the Rolls, chaired a wages investigation in 1942 that overruled the coalowners to deliver a minimum wage for colliers and a flat 2s 6d (half-crown) increase per shift.

—a breach of the Mines Act—he was punished. The law takes its course, and I suppose justice was served, but for one man and his close friends there must never again be any mention of co-operation.

During another afternoon we had a group of men coming out from a small district over an hour before finishing time. These men had repeatedly asked for more trams to be sent to them as they could fill more coal. The Production Committee had been told on several occasions about the need for a better supply of empty trams for these men, yet the shortage was still not altered. The men numbered about a dozen colliers and were working in a big seam where they would fill a lot of coal. This final afternoon they all waited for nearly four hours without a tram coming to any of them. As a protest they packed up and went out. Had they sat down their pay would have been right, because they could have put in for loss through delay, but they argued they were there to fill coal, not to idle about.

After that trams became plentiful in that district, but on the next payday each man was a considerable amount of his money short as a penalty for coming out. One working on the minimum wage was cropped more than a pound for that brief period in which he had come out too soon. The men had broken the rule, I know, but they did it with the intention of increasing the output. That is another little problem which will take some settling.

Benjy was in a bad humour when I met him in the street. He was 'poshed up'—an event which would have been worth watching—and at his front door ready to go out when his load of coal arrived. He had been complaining about the delay in delivery, but it came at an awkward time. 'Jawch,' he said, 'that old haulier done it a' purpose, aye he did. Do you call a man like that a Christian?' Coal delivery is one of the events of our colliery streets, and I often watch a load being delivered. It seems that the coal must be slithered as near the front door as possible; often some slides into the passage. A dust cloud lifts up to the mix with its kindred above, and the haulier makes a long hunt for the weigh slip, which he tries not to produce until the woman has found her purse and decided on a tip. Those who do tip are told he has brought a specially good load, 'all lumps,' even when a lot of small is evident, and those who have no tip are beneath his conversation. In many cases the coal has to be carried through the house to the back, and usually several neighbours join in carrying, the women struggling with lumps and jabbing shovels as if they too had been miners. The small children help joyfully, getting as black as tinkers and being especially busy with buckets and spades which were once useful on the seashore. The miner comes home, notices the black area outside his door where the coal has been swept up, and professes a disappointment, which is not very sincere, that the coal has not been allowed to wait his strong arm carrying. Proudly his wife takes him out the back to note how neatly it has been stacked, and usually he makes some complaint about the plan of placing lumps as being against all colliers' traditions. Then the atmosphere becomes sulphurous.

CHAPTER FOUR

Say half a mile from the baths to the colliery mouth and then another two miles of awkward going in heavy boots with lamps and other encumbrances. That takes the edge off your energy, and we usually have about five minutes of a sitting spell before we start work. That is the period when anything which has newly happened to affect our living comes in for debate. Often Benjy hears something on the radio or is told of a newspaper report, and by the time he has repeated this to us there is such a confusion that the news frequently carries an almost opposite meaning to its first report. Yet Benjy is insistent 'that way it was' and that way he will believe it. Horner said this, or Ebby Edwards said that, and 'By damn. They ought for to be ashamed of themselves.'[5]

Knowing our leaders better and having faith in them I refuse to believe Benjy's tale until I get confirmation, but as he has the start his version gets round before mine, and takes some contradicting. 'And you listen to me now,' Benjy insists, 'that's the way it was said. A feller was telling me …' Some mythical fellow has always told Benjy his mysterious messages. You can rarely get him more definite than that. In the past Hermit has been the sole supporter of Benjy's alarmist rumours. Hermit had a dislike for the Federation and especially for paying a weekly sixpence. Also the Union had checked a few doubtful activities of this mate of ours, and usually I had been the chosen weapon for that checking.

So between Hermit and myself there is a kind of armed neutrality. We are on different sides, yet we are not enemies. I study Hermit in an effort to discover why he acts as he does, and he watches me carefully in case I try to work anything over on him; in fact that is his fear of everybody. There are very few long-lasting quarrels in the mine, for men seem to get along very well in large numbers. It seems that two women can rarely work amicably, even in the same house, but a thousand men get along without any real friction in the mine. Of course when things go badly wrong they can blame it on the Government or the coal owners—many insist those are but different names for the same people—or on the management. Such a wide choice of scapegoats may explain why miners unite so closely. That feeling extended into village life before transport became so easy and men worked far away from their homes. Now their work and their homes are two different places instead of almost merging into one.

---

[5] Miners' leaders: Arthur Horner, Communist and President of the South Wales Miners' Federation; Ebenezer Edwards, Secretary of the Miners' Federation of Great Britain and member of the wartime Coal Production Council.

I can recall the period when, if a general meeting of all the workmen was needed, a crier with a bell would do the job in an hour. Now you must allow for the sending of many telephone messages and the arrival of various trains and buses. To avoid that we call as many pithead meetings as we think are needed.

No new arrangement, no decision, and no stoppage from pay is started until the works committee have considered it, made a recommendation, and had it supported by a vote of their men. Before this one shift we held a meeting, and when we got inside Hermit began to question about this shilling which was to be stopped from his pay. These details had been well explained at the meeting. After a day of broken sleep and a journey to work, during which everyone wanted to ask questions, I was feeling a bit edgy, and Hermit has proclaimed several times that he does not care who suffers whilst his end is all right. In work and out of it he has often worked on that principle, and that type of man has no welcome from me. After a couple of curt answers Hermit's mate brought in a complaint about the slackness of the Pit Production Committee. Having mates on that committee and knowing his statement was not true I snapped back that he would help them more if he came to work regular. Rude, I admit, but he was like most fault-finders—a very poor doer. Although he seems healthy I have not known him work many full weeks since war began. He subsided into the shadows, and his end was silence from then on; but Hermit still had something on his restricted mind and he kept asking questions about this new deduction.

'It was all explained at the meeting,' I said, 'and you had the chance to put any objections then.'

'Didn't like to ask, see mun,' Hermit explained.

'You were afraid to ask,' I stated. 'You like to make complaints about men behind their back.'

It was a true estimate of Hermit's methods, but even truth is something offensive, and Hermit found no pleasure in this bit. We forgot the passing of time in our enthusiasm at describing one another's shortcomings, and there was no politeness wasted for a few lively minutes. We were about evenly matched physically, and each had plenty of stored-up energy, so things looked ominous. The dull flame of an oil lamp showed the fireman coming closer to give us orders, but when he heard the commotion he suddenly remembered some place he had not inspected and went that way swiftly. George came ponderously but definitely to my support, and Benjy thought it was his duty to pour water on this wordy fire, but we sensed oil in his appeal instead, and things spurted anew. Benjy trotted off hurriedly, then things simmered down. At food time things were normal again. I was late at the end of that shift by myself, finishing a piece of work, and farther back I saw Hermit's light waiting. 'Case you might want help, see mun.' He had a peace offering; a nice block of sawn wood. We walked outwards together and then, awkwardly and incoherently, he tried to explain.

'You see, butty, I've allus been a sort of blasted mug, like. The old girl used ter tell me about that afore she died. Pretty clever she was, reading a lot of books. And now the two kids is getting a good age and they reads a lot too.

Ain't near so blasted dull as I am, see? Makes a chap feel out of it, aye it do.'

So obviously sincere was he that I felt about for a solution and suggested he too would find pleasure in reading.

'Aye, that's it. That's what I wanted to ask about. Where could I get some of these here books?'

I named a bookshop and suggested *The Ragged-Trousered Philanthropists*, which I knew was out as a Penguin and was simply written. Then I spoke of Upton Sinclair and Jack London. I explained the difference between a serious novel and one of the 'hug me, sugar' romances. Hermit listened, but was dubious; then he decided.

'Tell you what, mun,' he suggested, 'I got a quid or so I can spare. What if I brings it with me and you buys some books for me? P'raps you'll explain about 'em to me. Damn it. I'm gone I can hardly write me own name. Makes a bloke ashamed when his kids talks about things he don't understand, aye it do.'

I got him the cheap edition of Tressell's book, Sinclair's *Manassas*, Zola's *Germinal*, and also *An Architect of Nature*, in the Thinker's Library.[6] I supposed that Thinker dust cover would make him feel elevated; anyway it is a lovely book. I packed them and tied them in a bundle because I was sure he would never be seen carrying them loose. Blushing like a big boy who has been given the charge of a baby, he took them from me in the canteen and hurried towards the door. 'Hey, what you giving him?' Steve shouted. 'A new suit?'

'Yes,' I replied. 'And no coupons needed.'

In a way it was a new suit, for a subtle difference was about Hermit after a few days. He always waited patiently for me at the end of the shift and had a cup of tea ready on an isolated table in the canteen. He told me what page he had reached, what the characters were doing and what words puzzled him.

In his mining work he was a craftsman, and he began to take me to view his work. He was proud of its neatness and careful of his well-sharpened hatchet and properly set saws. Yet, although not more than middle-aged, he was beginning to lose weight, and he was tiring more easily. The mountain was beginning to conquer him. I wondered if there was something about that realisation which was influencing his desire for knowledge.

I have noticed it so often in the men who are beginning to age. They may have been twenty-five or thirty years underground, and if they had been in some jobs would have qualified for a pension. In the mines they feel their strength going and the limbs stiffening. The fear creeps into their consciences that their best days are gone; that even all their experience will not counterbalance their weakening body. They guess their wage packet will get steadily smaller, and they have no reserves put away. It is a grieving realisation, and must cause a deal of the bitterness which sours mining life nowadays. No hope for the future, no security for the family. The only prospect, when he cannot answer the call for work, is the miner's pension—parish relief.

When their working days are past, what can the majority of them do? Most had no interest outside their

---

[6] *The Ragged-Trousered Philanthropists* by Robert Tressell, 1914; *Manassas* by Upton Sinclair 1904; *Germinal* by Émile Zola, 1885; *An Architect of Nature* by Luther Burbank, 1939.

work; when that fails them they are adrift from all their connections. They know not what to do with their hands or their minds. The pit wheels revolve, the dust cloud rises, the loaded trucks are taken away and they have no share in its doing. Nor can they do any other work. I feel that every man should have some training in a second occupation that he could take to when his greatest strength fails. I know that few light jobs will ever be available at collieries, but would it not be possible to fit them for a factory job after, say twenty years in a mine? That is a long enough slice from a man's lifetime to be spent away from sun and daylight, in frequent danger and breathing forced air. Anyway, they should never be allowed to feel unwanted.

Neath is about ten miles from here. It is a fresh, clean town of considerable size. I think it easily the pleasantest of the Welsh mining towns. Possibly this is because I first visited it on a lovely day, and towns are like people—much nicer when the sun shines on them. Nor are there any collieries in Neath itself, but it acts as the hub of a wheel around which the life of the mining valleys circles. There we travel if we need some extra shopping or something which cannot be seen or sought in our smaller towns. I had got away from the centre and was walking down one of the working, yes … there again. Why do we always refer to the poorest sections of a town as the working-class district? Is it to be the accepted fact that the workers are entitled only to the poorest houses and the meanest streets?

Anyway, there I noticed a youngish man walking along the pavement. Recognising him, I stopped. He lifted his eyes from the pavement and knew me. Not having seen him in work lately, I asked what he had been doing.

'Hundred per cent,' he answered bitterly; meaning that he had been certified as fully disabled by dust.

'How are you getting along?' I inquired.

'They gives us two pun fifteen a week,' he said.

'If you could get something light to do with that,' I said, knowing well the futility of the suggestion.

'Aye,' he agreed. 'I hates this walking about with nothing to do.'

He had been to the Labour Exchange and tried two jobs. The first was carrying bricks up a ladder, and his tight-gripped chest would not stand that. The second had been easy on the first two days, but had changed to mixing cement on the third. A few hours at that finished him. Untrained, what outside work could he do in that enfeebled condition? And there appears to be a fixed idea amongst the men and the Exchange officials: once a manual worker always a manual worker. Nor does the urgent, rushing pace of mining work help much to accommodate a man to the more leisurely pace of surface work.

By one of the coincidences which are supposed not to happen I changed from the bus at Resolven and, hurrying to catch another bus, I met this man's brother Morgan. He again was completely disabled by the same disease. Pleasure at seeing me showed in the flush of his hollow cheeks, but I had to hurry as the other bus was waiting. I felt ashamed of this abruptness, for he might have thought that an old mate had no time for his pal who was in trouble. Probably it was some point about his compensation, but that was the last bus. It hurts to see that beaten look on the

W. I., Totally Disabled by Silicosis, c. 1943–5, pencil, 35 × 28

Dust disease is a repeated theme in the text of *Miners Day* and Isabel met several miners with pneumoconiosis whose portraits she drew. The disease itself could not be depicted, but her portraits capture the loss, indignity and tragedy that were its legacy.

features of a man who was once strong and independent. They were three brothers, all dust-laden, but the other had got a lump-sum payment of his compensation, and had solved his problem by becoming a publican. He had as good a claim as any retired police sergeant, anyway.

We had another pithead meeting to explain about the holiday pay. It has been raised to five guineas without any of the penalties which caused so much conflict before. In the emergency I acted as chairman. Just below I watched Hermit's eager face as we explained the agreement. He was so urgent and restless, I guessed something would happen. That suspicion soon became a fact. Most anxious to justify himself in my opinion, he was impatiently waiting for some point about which he dared to ask a question. When the chance came he did not wait for question time, but spurted it out with the vicious snap of a Japanese sniper shooting from an ambush.

'Will we chaps have to pay tax on that holiday pay?' he asked.

The words came out in one jumbled rush. Of course he would, we explained. In any case that money would be easier got than most which was taxable.

'Aw hell!' He was disgusted. 'I knew there was a snag in it somewhere.' The crowd laughed, and I reflected that now things would really happen. Hermit had asked his question and he had brought a laugh. Now he would have to live up to that estimate. So it proved, for he was at my side tightly from then on, a jaunty Hermit, strong with a new confidence. At the close of this meeting a reference was made to the framing of a four years' agreement.

A resentful voice from the rear said something which hung in my memory for a long while.

'What the hell is the use of talking about four years' agreements?' this man asked. 'By the way men are dying off from dust here there won't be many of us left in four years' time.' It ended the discussion, for many of us felt there was a deal of truth in it. During a long while there had been an average of two a week stopped by the doctors because of chest troubles alone. Take those figures along for four years and it certainly sounded ominous. I am reminded of another pit meeting at which a levy for a widow of a man killed in the pit was discussed. We all agreed, knowing that a period would elapse when the family would be short of funds, until the haggling over the compensation amount was settled. Then someone moved that we also make a similar levy for the widow of any man who died from dust disease. At the end a loud and decisive voice from the back insisted: 'Aye. And give it to the widow whilst the man's alive. Not wait until he's dead.' The sentiment was all right, but carrying it out would have been difficult.

Benjy has no time to waste with meetings. 'Like a lot of old women a' arguing. That's what they are,' he says. He had gone on inside, thinking to have a nice spell until the others came, but Nemesis in the shape of the fireman had found him, and his work was planned ready. He was drooping along, dragging his shovel behind when we saw him.

'Seen the fireman?' He answered our inquiry. 'Aye, I seen the nuisance. Wasn't a minute afore he gave me my blasted destructions. No need for you to larf. He got plenty of destructions waiting for you. You bet he have.'

Steve does a bit of mimicking. 'Fellow-workmen, I move that … all haw haw and BBC we was, but no half-crown for the riders. What we wants to know is when is that half-crown coming? Are you sure the blokes haven't been on the randy with it?' Not the least use getting ratty with Steve or the rest of them. That would start them on a tease which would be intensified every time we met. I just grin and wait for my turn to get a dig back. I know well that whatever they may say about their union, they will obey a decision properly taken, no matter what the cost. Starvation, victimisation, or jail, whatever the threat, their loyalty to their mates in a time of crisis would transcend them all. The trouble with our union, according to Benjy, is that we are too slow. He says, 'We should grab the iron when it is hot.' I know our hands are hard, but feel that might be a rather painful method.

George is beginning to notice and compare things. That was the first miners' meeting he had attended.

'Jolly keen, too, I say. Been to a few meetings in my time. Trades Councils and shareholders' do's, but they were tame compared to that lot. Shouldn't like to handle a lot of that sort of men. I'm beginning to wonder if they'll ever be satisfied. About that holiday now'—he starts to compare it with the lot of a small trader who would have no guaranteed wage and no paid holiday. I explain that until a few years ago there was no paid holiday for miners, and no apparent hope of getting one.

'I agree,' says George, 'yet they grouse like the dickens. I'm inclined to think that they will never be satisfied.'

I explain that the insistent grousers are only a minority, and we must harden our thoughts against these. They are, of course, usually the noisiest, but the sensible men provide an anchor at critical times. I know we have gained a great deal during recent years. I can easily recall walking miles to and from work in all weathers and spending hours in pit clothes, which were soaking wet. In some collieries men never went home dry. Going home once with every bit of my clothing—even to my singlet—sopping wet, I was so tired that I sat on a stone wall for a short rest. I must have drooped off to sleep at once, and it was freezing hard. Another passing miner disturbed me, but it was some minutes before I could move or speak. I believe that a little longer on that wall would have been the end of me. I tottered on in clothes that had been frozen stiff.

Many men are still eleven hours every day away from their homes, but now they have warm buses or trains to carry them. We have, too, that wonderful convenience, the pithead baths, to clean our bodies and allow a change into warm, dry clothes before we start. Here again human nature can be perverse, and some men will not use the baths. One made a grand complaint that he had found a flea on his shirt after bathing. From what I knew of his home conditions he should have been inured to more vicious things than fleas, but it seemed a way of impressing the other men with his spotless cleanliness. Again, with the food at the canteen, which has become more plentiful and varied during recent months. Be sure that the man who professes contempt at the cooking and serving of the meals there has to endure much rougher conditions at home; yet he must parade his epicurean tastes and his insistence on meals being served in first-class conditions. This attitude hurts the ones who try to cater for their fellows, and if they do not become thick skinned they give up their interest. This is a pity, for it often drives a man away from a job where he has been doing a deal of good.

Workers are undoubtedly the worst employers of all. They need to learn the way of treating their own employees. I am positive that no servant wishes more fervently to throw his job in the face of a harsh employer than do many of the trade union representatives long for a chance to finish with the carping, unfairly criticising members with whom they have to deal. All faults and no praise make a man either bitter or indifferent.

During several evenings I have been at the home of a miners' agent. He certainly has not to swallow pit dust every day, but he has other trials. Apparently he has no right to any spare time—no hour when he could claim that he had finished. All through the evenings he had caller after caller, and most of them did not fail to let him know it was their coppers which kept him in his job. If he failed to win a claim—no matter how badly founded—they let him know all about it. No doubt there were occasionally gleams of gratitude which helped to encourage him, and possibly it was long worry and nervousness about the outcome that caused some of the men to be so crude and overbearing in their manner. Skins get harder by such kicking, I expect, but if an agent thinks to please all his members he might as well get in a good store of arsenic right away.

The adjustment of the showers in our bathing is a clever contrivance. By the same lever we can alter the heat from ice cold to very warm; it is just a matter of slow turning. Usually I am a bit late arriving, and Steve has about finished. I stand by his cubicle whilst he finishes towelling, and he assures me the water is nicely warm, although he knows he has turned it to ice cold for the final shower—to ward off any danger of a chill when going outside. I step under and the coldness stops my breath whilst Steve shoves his hands against me to stop my backward escape. By the time I have swung the lever over he has made a dive, towels and all, for the dressing room, but I sometimes catch him and daub my black hands over his body. He knows better than to come back for a second shower.

Benjy is a queer sight in the bathing room. He looks like a boy with a bald head and a hairy chin. He has an accomplice called Job, who has a similar lack of size, but a great admiration for his own muscles and the confidence that a world's boxing champion was missed when they neglected him. He contorts himself to display his muscles whilst some of the men group around to express their admiration. I come along and insist that I see no sign of muscles, at which Job jerks himself afresh and keeps calculating the direction and hardness of my chin. Job had one leg badly damaged in the last war, and the other one smashed in a mining accident, so he certainly must have some pluck to drag himself to work even with the continual shadow boxing which is his second nature. Stark naked, Benjy and Job have a little spar before they start to bath.

They look like two flyweights contending for their old-age pension. The contrast is that Benjy retires defeated if he receives the first punch, and Job pretends to enjoy being hit—which pretence gives all the other bathers an excuse for adding to his pleasure.

One of the mine officials has a sense of humour and it helps him. Once he told me about a miner who threatened to 'clout' him. 'No bigger than me and seven stone odd,' he stated, 'and so a lot of blasted damage the two of us would have done.' Another time four of us were arguing. He turned a wicked eye as he passed, then doffed his hat and walked right into the centre of the dispute.

'Dear brethren,' he mimicked, 'may I add my testimony before we take the collection.' Another time, when the meat pies in our canteen were simmering under a bad reputation, he saw one miner buying half-a-dozen. 'Here,' he questioned, 'how many dogs have you got there?'

Benjy was ill one Friday night—really ill. He should not have been in work, but we knew why he came. Had he stopped home his bonus shift for that week would have been cropped. That one lost shift would have meant two; he would have been paid four days; by working that one he would be paid six. Money is short with Benjy, so he struggled to work. We knew the conditions, so in turn we tried to help him by unloading the rubbish for him. As we were in a cold part, and Benjy was shivering, he found an old coat and wrapped it around his shoulders while he sat down to watch us eat; he refused any food for himself. As he was walking away from the stones on which we had been sitting at food, Benjy met an overman who fancies himself as a wit. Seeing this unheard-of thing, a man wearing an overcoat underground in working hours, the official asked acidly:

'What's the matter here? Is it raining?'

'Raining?' replied the surprised Benjy through chattering teeth. 'Aye, by damn it is. April showers, sure to be.'

It had sure enough been raining, but stones were the downpour. A week's hard rain would have done good in that dusty district, and it was the end of April, so Benjy was not far out in any case, but Steve arrived and saw the need to support his sick fellow workman by adding: 'What a bloke wants here is an umbrella—like some folks take to chapel on Sunday.' This was a reflection on the safety conditions and on his habits, so the official thought it wisest to digest it silently and in the distance. Another time, old Crush, another rider intimate of mine, had hurt his leg quite a bit, but the official offered the advice that he always found it better to work through an injury, as it got cured much quicker that way. He claimed he had once fractured his skull and worked until it was better.

'Until it got better?' questioned the unbelieving Crush.

'Yes. I worked all the time until it was right again.'

'Huh.' Crush was still dubious. 'Are you sure that head have got better?'

And that devastating consolation of Dick Simons, who was busy cutting room for a post in the side and so was only half listening to a complaint by a new and harassed official about the stupidity of some miners under his charge.

'I assure you, Simons, that I have almost had to get

hold of those fellows and lead them by the hand.'

'That's how it is, bachgenny,' agreed Dick kindly. 'That's how it always is. It's the blind that leads the blind.'

And each evening another man calls to me when we pass in the mine on our way to work. His greeting is 'Can you drove?' This is a memory of a time, many years ago, when both of us worked together at a colliery in the other valley. In charge of the shift was an old man whose ability to express himself was his only weak spot. He wanted to make sure whether any newcomer could drive a horse underground; hence his first question of 'Can you drove?' We soon got to know the meaning, for as he informed us proudly, 'I can say the two spokes'—the spokes being the Welsh and English languages.

What a mixture of humanity is the average mine! We have the Welsh and English language merging and being used alternately, or sometimes half and half. Their nationals never quarrel about their separate languages. Also we have some Italians working with us. They have been in this country for many years and are good men and fine workers. War has not lessened our friendship with them. Also we have a Pole and have known American miners, French miners, Belgian miners. All have worked amicably alongside me: men who have started at most of the trades in the country, and who speak in most of the dialects this land has created. Yet gradually their work here nurtures in them two dominant features—the comradeship of men who live their dangerous lives together and loyalty to their great union.

And what a complication is the average mine!

In one section you have horses sweating when they are standing still and a tropical heat over everything: a dry gasping heat in which the water gets hot in the drinking jacks and each handful of small coal is warm. In another section of the same mine you have a ventilation so powerful and fresh that a man who is not working hard is shivering. It is quite possible for a man to be shifted back and forth each day between these districts, or even to work in both types during the same shift—that is one way to get a chill. Then you can have men in one section wondering how they can keep the roof up because it starts to trickle through any unprotected hole which is more than six inches square; while in another part of that same section you may have men using all their wits, pneumatic borers, and a load of dynamite to blow the roof down. You may have a collier who finds his coal working 'like a tide.' He has only to tap about his stall for enough coal to come free to last him his shift; while not far away another man surveys disgustedly a block of coal which seems cemented into its bed, and he has to pound hard all day to win as much coal as the next collier gets comfortably in an hour. Then you have the area where dust rises roof-high after every movement, yet in another part of the same colliery you may have men working up to their knees in water and not enjoying it a bit.

There appears a chance sometimes to use this surplus water to dampen the dust, but usually there are great difficulties to overcome. Often the water is so coated with minerals that it eats away anything it touches. If there are not solid blocks of coal or rock in the way, there will probably be a very long distance—two miles and more possibly—between the place where the water is and the part where it is so badly needed. It would need that length of pipes to carry it, and pipes need room. This last is not often plentiful in the underground roadways, and cutting extra width is a slow, expensive process. Usually it is better to bring fresh water in from the outside of the mine, and that is being done in some cases, where the dust problem has compelled it. A scheme for water infusion is being tried, and so is the method of spraying the cuttings of the machine coal. Sprays could be fitted along all the mainways and a fine jet of water thrown into the air current. This should have a beneficial effect.

The usual method is to pump a pool of water on one of the roadways and the walking men gradually carry this water onwards with their boots. Sometimes you get the water standing so high back on the roads that the journey rider gets wet up to his knees and demands that he be paid extra for working in water, while far on in the coal faces the dust is terrible.

It may be a long way in to one district, and to ease the men and save walking time you ultimately force the company to bring the men out in a journey of empty trams with a double chain for safety. That suits the colliers and other similar workers nicely. There will be no complaint from the boys either—but someone must be responsible for seeing that the couplings and ropes are secure, that the safety chain is fastened, that loaded trams are in front to keep the pull true; that the signal wires are right, and that those signals are properly given. The usual rider on that set will have to do these things. It will mean some extra work, so payment for that must be debated. While you are at it you might as well deal with the claim of the engine driver, who is kept a little late by drawing these men back and putting his engine right afterwards; and then again, far behind, are half-a-dozen hauliers who have to take their horses to the stable underground, but cannot start until that journey has reached safety. They will be latest of all, so something will have to be done about their claims—and so that little splash idea of giving the men a ride back has widened into quite a large ripple.

Just farther down the valley you may have another colliery working the same seams, yet meeting different conditions and having to make other arrangements. So there it is amongst the two thousand and more mines in the country in which over three-quarters of a million men work. Each one of that three-quarters of a million wonders how any new wage award is going to affect the industry, but more particularly how it is going to affect him. Behind that thought are the wives and families of the men, and a little further back the shopkeepers and professional men, who get their livings by and from the mining communities. Then also interested are the railwaymen who give this coal its first entry into the world away from our valleys. On a production of about two hundred million tons a year it brings a slice of work into their industry. Again there are thousands of buses with their drivers and conductresses, busy with the job of taking the miner or his folk back and forth.

It was over these many complications that the Porter award fell into rough weather. It gave a five-pound minimum to all underground workers and definitely benefited the lowest-paid men, but a man whose earnings brought him near the five pounds gained nothing at all. He was paid extra for some responsibility, but that extra was merged into the five pounds. Then on the question of house coal there were districts in Britain where miners had their coal for nothing, others paid three-and-six a week, and in other districts it went up to six shillings or more. I pay about fifteen shillings for a load delivered. In trying to place a fair balance all these different amounts had to be considered, as well as the different values of the coal given. Finally it was assessed at six shillings a month in addition to the usual payments. That meant that a man who was only part holder of a house and could have, say, no more than four loads a year, had three pounds eighteen more to pay for his coal. Many companies stopped this eighteenpence indiscriminately without caring whether a man ever had coal. Most of the trouble started, as I saw it, because the colliery companies insisted that the Porter award was the maximum payment as well as the minimum, whilst the men felt their leaders were at fault for not disclosing that this angle was opened with the Greene award, and no action was taken during the long period which intervened.

    Benjy is still not contented with his pay docket. The amount of tax is the topic at present. 'Aye indeed. I must be paying supper tax by the look of this docket.'

    Thinking of supper instead of super recalls the apology for his shortness. 'Jawch. I'm so blinkin' short as I've got to stand on the fender afore I can see what's cooking in the oven.'

Rhondda, c. 1943–5, pencil, 22 × 20

This street switch-backing along a hillside with monotonous terraces opening directly to the pavements on either side echoes many Valleys towns. It may be Brondeg Street in Tylorstown. The absence of windows on the right is to suggest the anonymity of such streets rather than represent reality.

CHAPTER FIVE

'These here are my brothers,' Tom Evans stated, taking that foully smelling pipe from his mouth and waving it to include the two men who were with him. Steve gives it as his opinion that Tom smokes oilcloth in that pipe and gets that oilcloth from the ashbins.

'This one got a shop of his own and this one is a vicar or something up Birmingham way,' Tom concluded the introduction.

'Curate,' corrected the miscalled clergyman.

'Aye. That's it.' Tom stood corrected.

'—to meet you,' concluded the curate, and we shook hands, watching the continual goings and comings of the men and boys into the baths and canteen.

'Most decidedly a picture of activity,' the curate agreed.

'It always is, especially on payday,' I explained.

'That's it,' Tom supported me. 'And me brothers is on holiday, so I brought them along to help me rise my pay.'

'Not to help you spend it?' I asked, guessing it would bring a contentious answer.

'Missus can do that easy enough.' Tom blew a mouthful of smoke up and seemed to watch his pay disappearing with it, and as easily. 'Besides, both of 'em gets more dough than me, and gets it softer, too.'

'Ah! Now! Now!' his clerical brother admonished him. 'We all have our troubles, and most certainly our calls on the purse.'

It is always a busy scene on that day, and our visitors are dustily impressed by one thing—the dust cloud that hovers around the colliery screens and village of Cwmgrach. They use the dry cleaning method on the coal—the water washery is the usual cleaning treatment—and as a result the air all around is laden with dust. There is as much need for a bath after fetching our pay as if we had done a day's work in the pit. Bright dresses are a useless extravagance, face cream only acts as a foundation for a grey dusty mask, and there is no need to make an excuse for a dirty shirtfront after the first few minutes.

A few thousand people live all their lives under the dust epidemic, hardly daring to open their doors or windows and trying to grow flowers or vegetables in a bedding that is pure coal dust which has been rejected from the sky. I understand the name of Cwmgrach means the Valley of the Witches. Frequently we call it something else—not mentionable in clerical company. Some of the old streets still remain hidden from the mountain winds up a narrow valley which breaks away from the Vale of Neath. Force of habit, I suppose, causes the occupants of these houses to mix a coat of whitewash with the grime on their houses

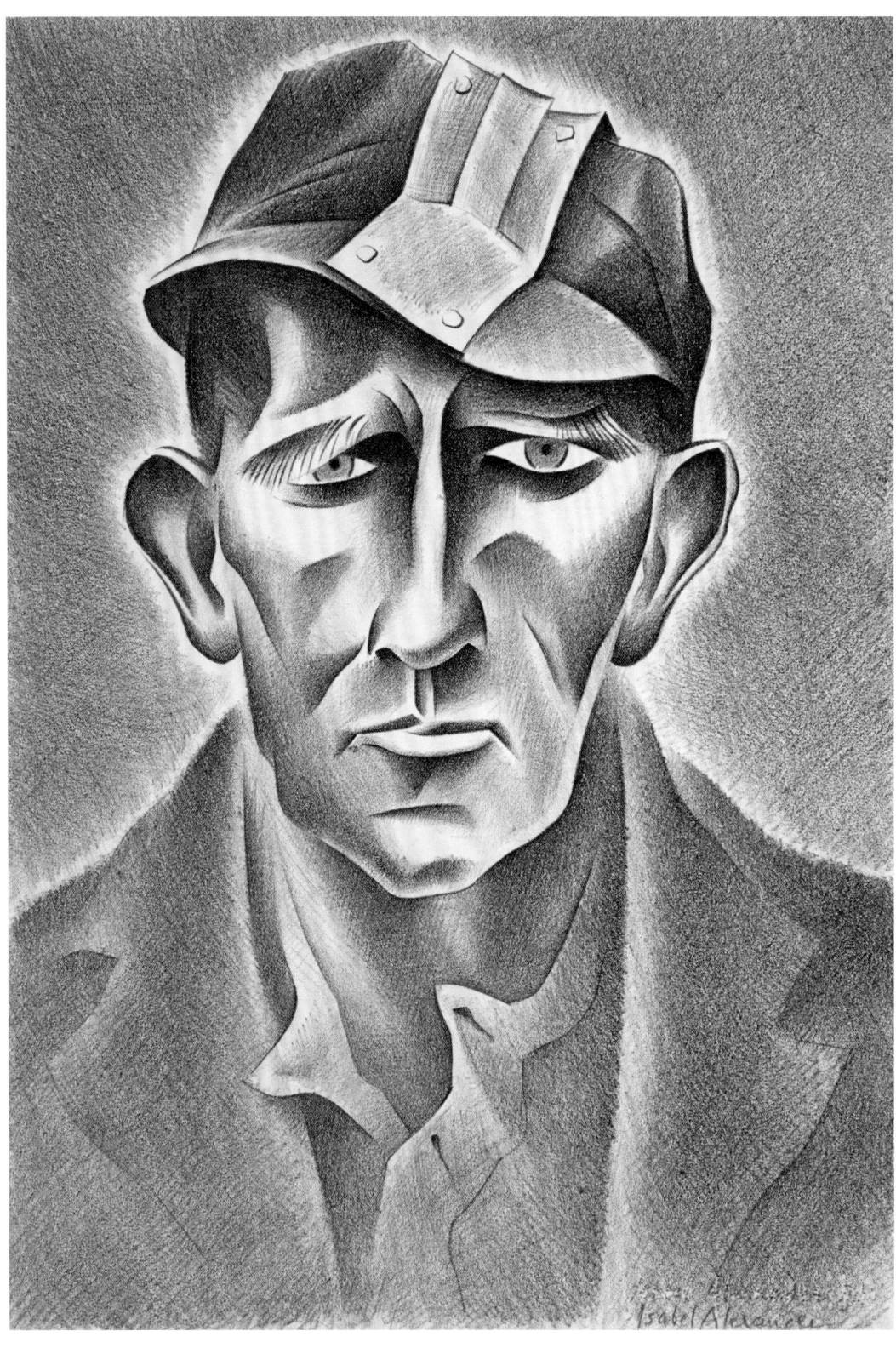

Tom, 1944, lithograph, 37 x 27

On the sketch (page 33) for this lithograph, which appeared in the first edition of *Miners Day*, Isabel wrote: 'This man was suffering from nystagmus, a condition showing rapid to and fro movements in the eyeballs. With miners it is brought on by working for a long time underground in a poor light.'

each year. Not much happens there, but as with most mining villages there are one or two of the type of my friend Benny—no connection with the other Benjy—who is forever organising lectures, concerts, or political meetings, so that the interest and knowledge of the inhabitants shall be stirred up. You meet a man of his type in every mining village, and they are always hard at work for somebody else's gain. They are always small in stature, large in heart, lean of face, and body full of energy.

As it was about half past two the day shift were rushing down the slope, black-faced and red-lipped, like Christy minstrels. The afternoon shift met them similarly dressed but white-faced. Little groups of men met to discuss their working conditions—and all over the area one could see a black-faced man and his white-faced mate sharing out the pay, puzzling over the figures on the docket, and planning their work for the coming days.

They sat about anywhere, confident that their clothes could get no dirtier, and pretty sure their sitting would add no more dust to the grass or the pavements. In a little cabin, shaped like a fowl-house, a queue was waiting to see the union secretary about some mistake in their pay. Another group waited on a high-walled platform to go in and draw their compensation money for that week. They were a battered assembly with arms in slings, crutches helping their legs, and often bandaged faces. As they grouped there, waving or shouting to the passers-by, I was reminded of the group outside the boxing booths at the old-time fairs. In this case it seemed that the various champions had been having a bad time—they had for their opponent the hardest puncher of all: falling rock.

Inside the canteen were more compensation cases waiting for the works policeman to signal that the last lot had been dealt with and their time was come. Inside the baths was another large group of disabled men being examined by the compensation doctors before they could go along to be paid their allowance.

It was only a pittance in any case, and many had come a good way to get it. Not so long ago they had to climb a narrow staircase to the top of an old building, but we insisted on that being stopped, and now the ambulance room at the baths is used.

Before the war the maximum weekly compensation was thirty shillings, and when we recall that the accident and disease figures during a period of ten years of mining were well over a million and a half, it seems that some poor devils must have gone very short, because they were unlucky enough to be injured. Since the war some supplementary payments have been made, and a single man at the highest rate now gets thirty-five shillings a week, and an additional five shillings after thirteen weeks' idleness. A married man would now get two pounds a week at the start, and another ten shillings after thirteen weeks. These are maximum payments for a man whose earnings entitle him to the highest rate. There is also an allowance given now of five shillings weekly for each child. These payments are subject to various conditions that they must not exceed two-thirds of his pre-accident earnings. There is another definite rule that no payment is made for the first three days lost unless he is absent a month from work. Knowing that rule many men, who feel they will only be off work for a day or so, do not bother to report an injury, and sometimes lose severely by their optimism, because each accident must be reported—if compensation is to be paid—before the man leaves the colliery.

On any ordinary day Tom would have had a drink or two with his cronies. Escorted by such respectability, however, he made a grimace towards the public house and we decided to have a cup of tea. The washed and brushed-up day shift was coming in from the baths. Three long tables were served with hot meals brought from a local British Restaurant,[7] and at the other men were drinking tea, eating meat pies, and arguing about films, football, girls, books and politics. A very animated scene, and the curate approved the friendliness and the cleanliness of the place, whilst the trader brother wetted his lips at the sight of a six-deep crowd by the counter and five attractive girls serving at a rush.

We sat down; our two companions preferred to stand. I have always noticed that in their leisure hours the manual workers look quickly for a seating place whilst the sedentary workers prefer to ease their muscles by standing. I noticed other things too, most especially the contrast between Tom and his brothers. One was two years older than Tom, the other three years younger, yet Tom looked almost old enough to be their father. His frame was drained of every ounce of fat, his hands were curved inwards like claws, and his shoulders had a stoop. Of course he had plenty of blue marks on his hands and face.

His brothers were wider around the waist than the chest; with Tom it was the reverse. He had dodged death a thousand times and had very often worked on after injuries which the others would have counted serious. One time he had continued to work and had hobbled out after what he termed a 'tap' on the foot. Next day we found two toes were fractured. He was indignant about the fuss made when he had to stay home for a few weeks. I wondered what would be the curatory reaction to such an injury.

Tom had been a collier for most of his life; a good filler working in a big seam. Let me put it at a low and easy average of four tons a day and only two hundred and fifty working days in a year. A thousand tons a year, and for twenty years, before he had to take on a lighter job in the mine—or rather, a less rushing job. Twenty thousand tons of coal—fifty big trainloads. That was part of Tom's contribution to our national wealth. Then we will take an average cutting price for such a big seam and call it two shillings a ton, so Tom was paid five thousand pounds for his coal-cutting over those twenty years. Of course he would have extra payment for setting up timber and other small jobs, but the coal would be the main item.

What energy, what comfort, and what wealth had been created by those fifty thousand tons that had fallen away from Tom's mandrel point? Yet Tom is just one of the thousands who work to ensure our national output of about two hundred million tons yearly. Last week we saw a cutting pinned on a post in the mine on which a scientific

---

[7] The British Restaurants were community feeding centres set up by the Ministry of Food in 1940. By 1943 they served 600,000 nutritious meals a day at low prices.

expert stated his opinion that there were still thirty thousand million tons left to be mined in the South Wales coalfield. Thirty thousand million tons! That left us astounded.

'Jawch,' complained Benjy, 'and I did think it was nearly all worked out.' The general opinion amongst us was that it seemed safe to buy a couple of new mandrels.

'S'long,' said Tom. 'See you in graft tonight.' The talk and the tea drinking had slackened, and by train and bus the workers were starting away for their homes.

'Very efficient and well conducted,' said the shopkeeper, who had either a sad or a thoughtful look. I feel he would have liked to take that canteen, or the girl assistants, with him—or perhaps both.

'A very good afternoon. God be with you.' The curate made a flourish of his hand shaking. Those hands of his were soft and white. I hated to think what Tom's crowbar, or the big sledge handle, or just a few stones dropping from the roof would do to their manicured beauty.

They hurried away, and walking between them Tom, crouching slightly like a boxer going to meet his opponent, had the appearance of a criminal being escorted by two detectives. In their young days, so Tom told me later, they were 'as alike as three peas.' What a quarter of a century of hard manual work can do to a man, even if he was as lucky as Tom!

The secretary had come from his cabin and was nibbling at a meat pie whilst he listened to a report from the chairman. Only two groups were left in the canteen with the width of the large room between them. The workmen's committee had collected near the door. The mine officials talked in low voices not far from the counter. The girls swept in and tidied up, shifting each group impartially. Looking sleek and well fed amongst the gaunt-featured men, the compensation doctors got into their cars and drove away. The compensation clerks folded up their books, the pay clerks collected the remaining envelopes and checked their sheets. The pit screens shivered and rattled as more coal rolled over them and more dust rose into the atmosphere. Children hurried home on this day of possible benevolence and 'pinked up' housewives with marketing bags queued up for the buses which would take them whither they judged their money would be best spent.

Their conference over, the mine officials moved out towards their homes, and with the girls urging them off so that they could close the doors, the dozen members of the workmen's committee strolled off towards a private room in the big building which housed the baths and the canteen. Whilst some belated bathers were still mingling a song with the soapy water they sat on the forms and began. 'Minutes of last meeting,' then later 'Correspondence,' and the secretary read through the pile of letters from Union headquarters, from Agent, from Blind Institute, about Hospital Governors, about new mine legislation, about post-mortems, inspector's reports, requests for something from the Artificial Limb Fund, Medical Board reports, compensation payments to some injured workmen, to elect a delegate for one or two conferences, to support this or that political resolution. The real business starts when it comes to workmen's complaints about working conditions or payments. So-and-so was not paid extra width and wants to claim it. A committeeman is picked to measure the distance next day.

Another collier sends in a complaint that coal was filled from his area and wants to be paid for that coal. What are the facts and what is the custom? Someone there has inside knowledge of both. The coal had to be cleared because a crush had started and it was urgent. A cog of crossed sticks was needed in place of that coal to steady the roof. The filler had not broken a custom as the cog could not be left until a proper collier came in on his shift. No case voted, and next day the men concerned will be told of that

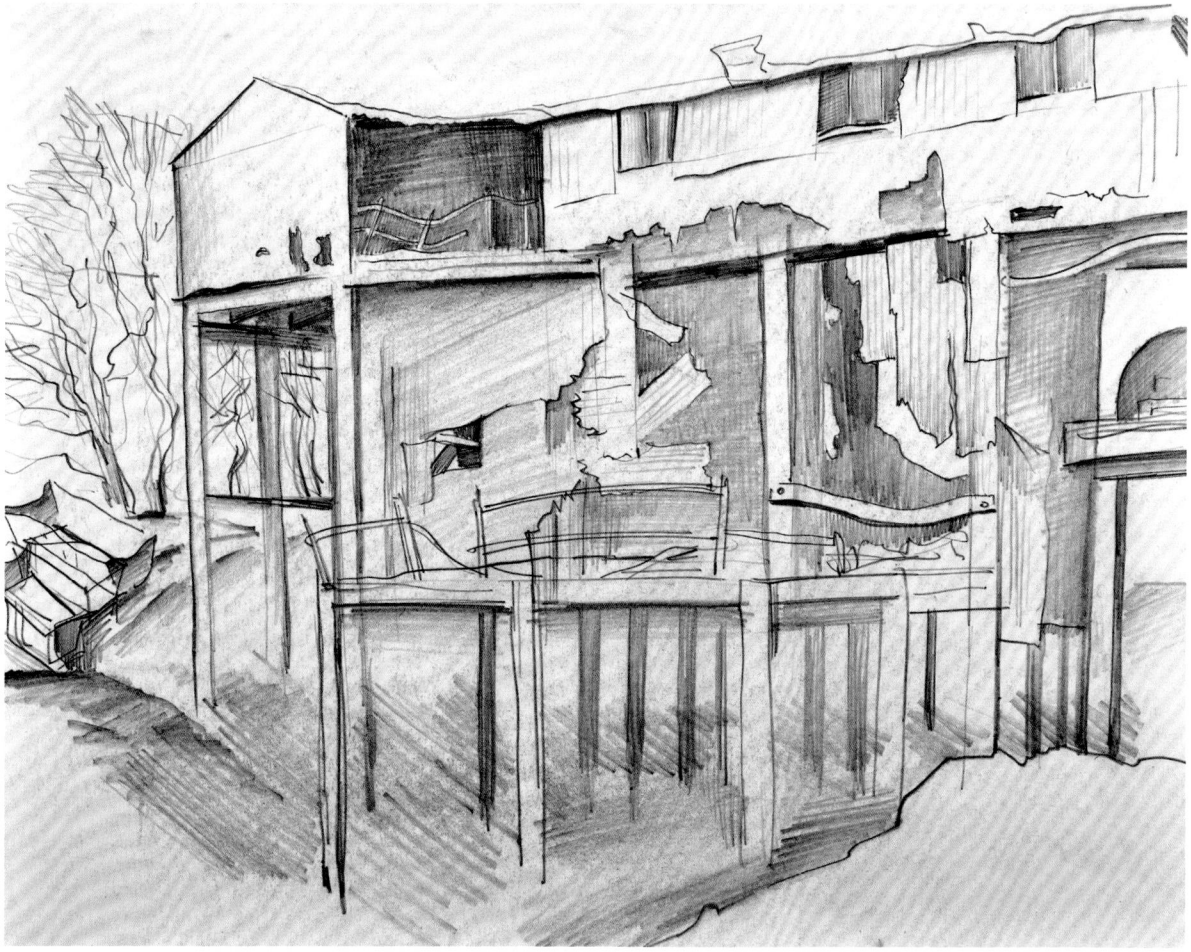

Screens, c. 1943–5, pencil, 28 × 36

The dereliction of this screens building for sorting coal and loading it into rail wagons underneath speaks of industrial decline. Corrugated-iron sheets have rotted away and railings have collapsed. This may be a disused screens near Glynneath where Bert was photographed walking by Bert Hardy in 1941.

decision. A youngster has been shifted and put to do a job at the conveyer end in place of a man. He is still given the boy's wage. They decide if he is doing a man's work he must have a man's wage. The management will be told of that next day. Any more complaints? A shoal of them are given and dealt with. Then the agenda for the weekly meeting with the management is prepared, after a report of the last meeting had been accepted. Then the pit Production Committee gives its report, and suggestions for increasing output and removing blockages are considered. They will be sent before the joint committee and possibly on to the Controller. What will be the ultimate fate? The meeting appears pessimistic over that. 'Any other business?' Always a lot of that, and then the meeting is closed.

'Anybody got the time?' Nearly seven o'clock. 'Good Lord, I must run.' They troop out into the fresh air past the bath attendant who is hosing the concrete floor. The treasurer taps his pocket to see the bankbook is safe, the secretary stuffs papers into his case. 'So long, boys. See you tomorrow.' The analysis of the mine business is over for another week, although problems will be handled each day. Such is the business of the men who are the core of Trade Unionism, and are the arteries of one of our great industries. A worrying, often thankless job, but a great place for a student of human nature and of mining life.

I wonder who took the trouble to pin that bit of the *News Chronicle* on the post near the double parting. Did he think to cheer us or warn us with that cutting which stated that deaths in the mines from fatal accidents were seven hundred and ten during 1943? Possibly he had the same motive as the evangelist who pushed a tract into Benjy's hand one evening. 'And he did give it as I was coming to work. See? Said "This will help you, my friend." Jawch. Been more help if he had come himself and helped me to chuck this muck out.'

Benjy complains that his food is not good enough for him to do a lot of hard work. 'It ain't got enough of them proto' things as they talks about on the wireless. Look at me tonight. Mermaids' grub; that's what I got.' Apparently that mermaids' food was sardine sandwiches. What proof Benjy has that mermaids spent their mealtimes gobbling up sardines I have never been able to discover. He is a little concerned also about the prevailing fear. 'Aye indeed. Sometimes I think I have got that pneumonia coaxes,' which is his version of Pneumoconiosis. Apparently that simple little word is the result of long and united medical research into a name which will cover all the dust diseases. It has nearly added another complaint to our numerous mining list—that of lockjaw.

Lovely flowers often grow from repulsive rootings, and it seems that way sometimes with human beings. I watch the children making their imitation sandcastles in small coal and building their stone-circled play homes on the ash tips where there are plenty of tins to take the place of crockery. In the warm days of summer they paddle in the river—once clean flowing, but now sullied with colliery waste. It seems the habit to build slag tips near the riverbanks so that when the water rises it sweeps part of the tip away and leaves a place for more to be tipped. I have not been told what harvest the fishermen get after

Tip Behind Penygraig, 1943–4, pencil, 19 × 23

A level worked at Gelli-faelog on the hilltop west of Penygraig in the late nineteenth century but was later the site of a tip for Naval Colliery. The conical tip is spreading over an older fan of spoil covered with thin grass. Power cables march across the mountain behind. A slightly different view appears on page 132.

casting their lines into such contaminated waters, but the children splash and yell with every indication of pleasure—although most of them would be dirtier after contact with that water than they were before. Of course, Glynneath, especially, the education authorities have done their best, for the schools are well kept and the newer ones are especially attractive—to grown-ups. The infants' school is a little gem; far away from traffic, with a great admittance of light and fresh air. It is surrounded by flowerbeds, which are most carefully tended and renewed. Not far away, at Pontneathvaughan, they have built another lovely school, well lighted and high above all dust. It is a memorial to Thomas Stephens, a Welsh historian, and, so I feel, a much wiser decision than a stone monument that has no other purpose than to reopen old wounds.[8]

This area is, for a mining valley, a great deal more attractive than most. Nature saw to the beauty of it, man put the villages, the works, and the slag tips to show how vile it could be. The mountains are impressive, with immense expanses of green and gentle slopes. Frequent piles of stones and path tracks show where a hill farm has been abandoned. Hardly half-a-dozen along those many miles of grazing land are still occupied. Mountain sheep, not much bigger than large rabbits and quite as swift of movement, nip at mountain grass and watch their chance to steal a good feed from some unguarded garden. High up the slopes the trees are twisted and bent over in continual obeisance to the wind which rushes across there on most days. If the sheep, and the folk, who live on those mountains are of similar calibre to the oak which insists on growing up there, then they must be tough indeed. The hatchet glances off this oak and the sparks fly as if steel had been sliced.

The waterfalls are amazing and plentiful. I know several with a sixty-foot leap, and as our rains are heavy and often we can usually have a look at a waterfall in real action. So all that is ugly in our area has been done by man.

---

[8] Stephens was a nineteenth-century historian, literary critic and social reformer. He was born at Pontneddfechan (Pontneathvaughan) and spent his adult life at Merthyr Tydfil. The school opened in 1930 and closed in 2008.

Looking at the children chattering away on their release from school I felt that surely no area could show more lovely children. Clean, healthy and finely featured, they were like a garden of talking flowers.

A little later I saw four bonny little children near the back door of a house. Knowing them very well I tried to find out why they were so subdued. Not learning much I asked where their father was, and was told he was away working. Where was their mother? She was in the house, and crying. I went inside, being used to doing that in their home. In fact I knew the early parts of their story. Their father was aged about thirty, a collier all his life until some months before, when his doctor ordered him not to work underground again. An X-ray examination by the miners' own radiologist confirmed the presence of dust and it was suggested, because he was young, that he had better not go underground again. Probably an older man would have been advised to go on working until he was completely finished. Then came a five months' wait for the sitting of a Medical Board. That decided he had a deal of dust, though not enough for even partial compensation, but advised him to find work on the surface. Six months'

Untitled,
c. 1943–5, pencil,
36 × 26

This intensity of this sketch of a young girl recalls some of Augustus John's drawings of his own children.

Glyn, 1943–4,
pencil, 28 × 23

A lightning sketch of a young boy. Isabel wrote 'portrait of Glyn' as though she might have known him well.

hunting amongst the large number of men similarly placed, failed to find him a job he could do. His wife went to work in a factory for a while, but, being not well and never strong, collapsed and had a period of helplessness during which he struggled to carry on the home and the children's care. Then he was sent away to an English town for training in a new job. Because of his age the Army authorities called him for a medical examination. They definitely contradicted the civilian Medical Board, and reported that the man was suffering incapacity from industrial disease and should not be far away from skilled care. He was brought back again for a repeat of the mining Medical Board, and they reversed their earlier decision. He will now get a payment of partial compensation, but had travelled back to factory work. How he struggled to work I do not know, yet I saw a letter from his new employer stating he was one of the best workers he had ever seen, and that he hoped to keep him. I also saw his pay paper, which showed that after paying his lodgings, even without taking any pocket money for himself, he had only thirty shillings to send home to his wife. Rent and living for five out of thirty shillings. No wonder she cried. She had been to the relieving officer with all the facts and had asked for help. She got none. If charity begins at home so, I feel, should humanity, and we need to ponder whether all the indifference and brutality in this age belongs to other nations. Fortunately I have some experience in getting justice done, and I lost no time in setting that knowledge to use.

I should add that they were living in one room up and one down. In that manner they had lived for ten years, never having anything different since they married. Yet the home and the children were very clean and well cared for. I know families that have waited twenty years for a

Mrs C., Mother of Seven, 1943, lithograph, 41 x 31

Isabel noted that Mrs C. was the mother of three of the children she drew and that the setting was near her home. It is a troubling portrait that goes beyond Isabel's habitual objectivity. In an otherwise warm review of an exhibition of Isabel's in *The Studio*, January 1952, W. J. Strachan picked out its 'almost brutal stylization'. However, the distress was real of mothers who had starved themselves to feed their children through the Depression. Seven children would have meant a battle for survival, and Isabel noted that Mrs C.'s husband was among the long-term unemployed. Her up-to-date hairstyle cannot diminish her gaunt frame, lost teeth and traumatised expression.

home of their own and about one-third of the houses in these valleys had not—nor had this home—any lavatory or bathroom, or even hot and cold water. If we are sincere in our ideas for better housing after this war we have a long way to go to redeem the neglect—and the promises—from the last war.

On Wednesdays many of the night shift and other men, who are temporarily off work, get along to Neath in the morning. These are usually the men who have come themselves, or with their parents, from country areas. Often breeches and leggings are seen on men who we never supposed wore such things; their normal garb being moleskin trousers and strapped yorks below the knees 'to keep the dust out of their eyes.'[9]

Walking sticks are the incitement for derision underground as being the symbol of an official, but the venturers towards Neath frequently display a stick and are proud of it. 'Made this 'un myself. Holly, it is.' Its chief handiness seems to be the prodding of some ruminative cow, as a pointer of cattle-handling skill. I have seen the same men, and often their women, at market after market. They never buy, probably they have no money to buy, and in any case they have no place to keep such livestock.

Yet week after week they crowd the buses, wait for the lorries to unload, pass judgement on young stock, and shake their heads after walking along the tied rows of cows and calves. They are waiting in a dense crowd when the selling starts and the real buyers have difficulty in getting near. They follow the actions and bids of any well-known buyer like film fans watching their stars, and often the dealers, well aware that this is their brief day for which they have donned the whipcord, swagger about in a majesty which is sometimes spoiled by getting at the wrong end of a cow at the wrong time.

The auctioneer spices the sale with his sayings. 'Now then, you Wednesday farmers, open out a bit and let the real farmers have a look.' 'She's a beauty,' insists one dirty-looking seller, leading a dejected cow forward. 'So she is,' the auctioneer agrees, 'and so is the owner. So is the auctioneer, too, only nobody will take any notice of him.' And I recall that sale of an old grandfather clock. 'Going for seven pound. Going. Going. Going. Gone. And that's more than he have done for five years.'

They pet the calves, and they appreciate the pigs, getting there in a mass through which the auctioneer has to force his way. Interest in pigs is more understandable as being nearer a more or less enjoyable possibility. As Crush says, and he is a regular visitor: 'If I had enough money like, I might buy a pig—if I had a cot and some grub to give him.'

Neath is a small market and selling only lasts about two hours. The visitors do not seem to bother about the town. They wait for the next bus home, to treasure for another week that short glimpse of the old life. The farmers remain behind to conclude their business and the market is almost deserted. What attracts those landless people to that weekly journey? I was born in the country, too, and what attracts me there, anyway?

---

[9] Many miners wore garter straps to stop trouser legs riding up when kneeling and keep out insects or vermin, which might bring tears to the eyes if they reached too far.

CHAPTER SIX

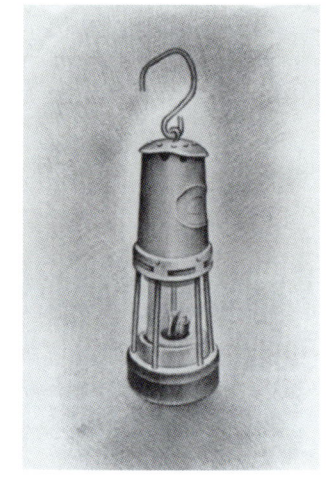

Miner's Lamp, 1944, lithograph, 13 × 8

An original illustration for *Miners Day* of an oil safety lamp of the kind that Bert used. The carrying hook and top are battered. The shape of the maker's plate indicates it is a 'Cambrian' lamp produced locally by E. Thomas and Williams of Aberdare.

Yes, it was once again a Friday evening. Just about the time when most of the afternoon workers think with delight of another underground week that is finished, whilst the night shift consider gravely that if they can get along until the next morning they will be due for a brief weekend spell, too. They too excite the envy of the day men who climb blinkingly up the tram incline next morning as the workers off the 'owls'' shift are going home. This late Saturday finish for the day men does cause a deal of complaint. Their ending is at the same hour as any other day, so it must be half past four, or later, when some of those who live at a distance get home. An earlier finish on a Saturday would make some—especially the younger section—of the day shift much pleasanter on that day.

I did not feel especially pleasant on that Friday evening either. That day is always a worrying one, with its problems about pay and its broken sleep. Also it was cold and raining, making the baths seem a desirable stopping place and the sticky, filthy incline the fit road to a place detested. A few men lingered outside amongst the shelters, but the majority had gone underground, away from the dreariness of that night in early May.

It was early enough, and I felt warlike. I had a grouse which had been simmering for some weeks, so I decided I would go up and see 'them.' I walked past the coal trams, climbed the stone steps, knocked at the door, then entered the cabin of the mine officials. I have long ago shed any awe of such surroundings, so I made myself comfortable. I picked a good seat on the plank bench, placed my twelve-and-a-half pound miner's lamp on the floor beside their two oil lamps, glanced unbelievingly at the stack of report books, and then started to explain my coming to the two over-officials who had remained behind after the fireman had gone underground. In past years I have acted as leading hand in many difficult mining jobs and in various mines. During recent years I have thought it wiser to act as a second-hand repairer, leaving most of the responsibility and the higher pay to other—sometimes men with less mining experience than myself. When in charge I had frequently worked myself to a standstill, so I judged that my working life would be longer if I took second place and, as a miners' committeeman, I knew well the need to prevent any blame being placed on my shoulders by the officials.

The two overmen had finished their books and were enjoying a last smoke before going into the mine. The room was bare, dusty, and very warm from the anthracite coal which had been taken off the many loaded trams standing just below. They prepared to listen to one of the ordinary complaints which come from the men's representatives to the management, but I reassured them that this was just a social call on a matter which concerned myself and my treatment. They sat up all ready to give attention.

These were my arguments. I worked in a new district where repairs were unusually difficult. Normally we had been two parties of repairers there, with an extra couple coming in when an emergency occurred. Very soon after Christmas both the higher-paid repairers were taken ill; we knew that even the first would be many weeks before he resumed work. At that period continual crushing made the repair work very urgent if the area was not to be closed up. Repairers were scarce and badly needed. Coal, also, was scarce and badly needed. We hear so many appeals for co-operation, and from that angle I agreed to take up again the repairing work until the other men came back. There were men with mining experience available to assist as second hand, but instead they sent a new starter underground to help me at that most difficult and dangerous of mining tasks. Several times during that period, on hearing that crushing was taking place, we had hurried in, thrown off our clothes, and rushed into the work all through the shift, with only a snatch at our food while standing up. At least three important places had been saved from falling in completely, apart from all the other work. Three working places, that were turning out at least six trams each a shift. The coal had come back without a break, the whole district had expanded, and had helped a great deal in at last passing our pit target and having the flag at top mast waving proudly over the baths. The emergency repairers had come there no more frequently than formerly.

I explained all this patiently to these two officials who should have known the happenings quite as well as myself. They got reports and paid frequent visits, so nothing prevented them knowing. I had waited during the last two months for some recognition of my efforts and the awakening of their conscience. As that conscience still slumbered, I explained, I was here to give it a jerk.

I inquired, after explaining how much they had benefited financially by the other men being absent, when they were going to pay me the extra rate given to the other repairers. Under the present conditions I was getting exactly the same pay as when I acted as second hand. My mate, with only a few weeks' mining experience, was getting equal pay with myself.

The Porter award is to blame for that, they said, and we are not responsible. That's an excuse, I objected, for the Porter award fixed the minimum pay only; you can always pay more. They argued that it was not their job to pay me, their duty was to get the work done—the maximum amount too, apparently. I argued that any man who had authority to make a man do work should surely have authority to see that he was paid fairly. That point did not seem to create any enthusiasm.

Then I came along on another angle. The colliers who train Bevin boys are allowed extra for that training. I, too, had a beginner with me, although he was middle-aged.

No one can argue but that repairing is more difficult than coal-cutting, and the need for a skilled assistant is correspondingly greater. When I was working at a height of perhaps twenty feet from the ground I had to climb down to shape a lid properly or trim a post in the right way. An experienced assistant would have saved me all that climbing. George was willing, but had not the years of experience which were needed. So I claimed that there should be some extra allowance for this added work and responsibility. They agreed that I ought to have much more pay than a beginner, but rejected my suggestion of how to alter it—by giving me more.

When the pained silence which follows that argument had become wearisome I attacked again. I argued that I had ample tools for my own work, but the jobs I had been doing lately needed a greater variety of tools. I had used those of my absent mate, and I knew he would not complain, but in that using roof falls had broken two sticks, the hatchet edge had become thick, and both saws needed sharpening and setting. I meant to get them all put in good shape before the others started back and it would cost about six shillings to do it. Was I to be compensated for that spending? That idea again brought me no help or encouragement. My complaint, so they assured me, would be passed on, vaguely, higher up. To someone who was probably sleeping the slumber of the just at that moment.

'OK,' I answered. 'I got just as much as I expected. Good evening.' To end with that episode, I had a message in a few days that no extra at all could be paid. So that night I did not hurry in, nor would I have even if the whole area had been crushing.

Far inside, I lingered to hear the main rope rider outlining his work for the night to his engine driver. 'There's ten of coal, eight of muck, two trolleys of timber and the water tank on the shunt parting. I'll pick that lot up and drop 'em on the chain by Grigg's Deep. Then I'll take fifteen empties down to the middle parting for the lower engine. Bring all that coal up, see, then we'll have time to shunt that other lot about a bit to get all the muck together.'

There is always method in these deliberations, as the engine driver can then gauge what is going on at the end of the steel rope which winds in and out over the drum in front of his face. In some places he must allow the rope to run out easily while in others he must keep his brake just touching. The rider, also, has his special method of signalling: sharp double knocking on the electrified wires if he wants a quick pull, and a long-held knock if he wants slow and cautious movement. The engine drivers can usually tell if a strange hand is holding the knocking file across the wires, and the rider will often be able to state whether a strange hand is on the engine throttle. To be sure of your driver means much in the confined roads of the mine, where there is often no safe place to jump off some badly swinging rope or the trams it pulls. Also the shunting of various materials is sometimes as complicated as in a railway assembly yard.

Farther down I followed the receding lights of other men as they went inwards. Finally I got my tools, and went on to our working place for that night. George was

Blaencwm Level, Waiting for the Train to Pass,
c. 1943, pencil, 28 × 23

This swift sketch in a hillside level at Blaencwm shows a miner crouching in a niche or refuge hole next to the tramway, waiting for a ride of trams to pass.

---

[10] Steel rods around a metre long with forked ends, used to prop upright the steel arches or paired 'arms' of timber while fixing them in place.

already there, filling a refuge hole which was constructed for three men and peeping apprehensively along the way I was supposed to come. 'Thought you had gone back home,' he greeted me with obvious relief.

'Just went up to have a confab with the big noises about some extra allowance,' I explained.

'No good, I suppose?' George had already learned the pessimism of his workmates.

'Not a sausage,' I agreed. 'I suppose we had best do a bit.' Later, as George had not come back from a journey after steel forks for holding the upright arms in position,[10] I went to look for him. He was in one of the stall roads staring at a huge lump of coal that blocked the complete width of the roadway.

'There's a whopper,' he said. 'What do you think of that lump?'

'Decent lump,' I agreed. 'Lucky there was no one just here when it rolled back. Seems that all of the slip came out.'

'That's a chunk out of that thirty thousand—what did it say on that cutting?'

'Thirty thousand million tons. Oh, there's plenty left yet. That piece would surely weigh about three tons,' I said.

'What the devil will they do with it now?'

'Break it small enough to lift. It'll mean a dose of sledge and wedging, or p'raps they'll get a pneumatic pick. Be much easier for them if it had come out in smaller lumps.'

George was fascinated by that huge lump of coal. I could sense the thoughts going through his mind. What if someone was under it? Our pit sense is always to make a swift examination to see if anyone is under something that has fresh fallen. Not a pleasant undertaking, but it has to be done, and swiftly. I remember an instance of a man caught by a smaller lump when passing through a narrow airway. He could not shout or pull himself free. Six inches past his reach was a rail which would have given him a handhold to drag himself to safety, but he failed to grab it, although he wore the skin off his fingers in the effort. Men passed within a few yards of him, yet he could not call to them. When he was found it was too late.

It seemed we were now working in an area of strong coal, and very thick slips, for the next night there was an almost exactly similar lump in the same stall. This one we had to break up, and George had a go. This type of coal is so strong that if you drive a sharp mandrel point into it the steel pointing will sometimes snap right off. The usual method is to make a small hole and drive a steel wedge in the place we judge weakest. We spit on the wedge point before driving—why, I have never found out. Then, with each blow of the sledge, it is the habit to make a 'hussing' sound in the style of an old-time ostler brushing his horse. There is quite a knack in hitting the wedge-head true in a feeble light, and George is nowhere near accomplishing it at present. His sledge hits everywhere and almost everything except what he intends.

Steve tries to be helpful when he arrives.

'I'll do my share, Georgey boy,' he says. 'Now, ready? You hit and I'll grunt for you.' George's hiss has deepened into a full-bellied grunt. Steve has also a playful way of dabbing his fingers swiftly on the spot and saying, 'Right there,' just as George starts to strike. George, alarmed lest he should hit Steve, and checked in the middle of his erratic stroke, then gets into all sorts of contortions. After one stroke the sledge came none too gently against his own ankle, with blasphemous results. Whilst he was rubbing it, Steve took a piece of chalk and whitened the top of the wedge.

'Hit the white mark,' he advised, 'not your ankle.'

Ponderously, like a charging elephant, George rushed at Steve, and the rider stepped back and flashed the full force of his electric lamp in George's face. Blinded, George grabbed at the darkness whilst Steve danced backwards and called, 'Next time, big boy. And Bob's your uncle.'

I wanted the hatchet off my tool bar and George hurried off to get it. A few seconds later he returned and silently unhanged his lamp, this time taking it with him. He has not yet found it second nature to take a light with him everywhere.

During one shift I had to take George along a return airway and down into a smaller seam. As we descended I explained the method of boring into the strata and blowing it down, the different strengths of the ground encountered, how they tested to see if all the loose pieces were down and how they kept the slope gradual.

These hard headings, as we term them, are the places where miners contract silicosis. This is a stone dust disease and the stone dust hardens into a cementy mass in the lungs, gradually making breathing impossible as the man's inside seems to be turning into a stone. Small coal dust, which causes anthracosis, has a different action. Its sharp particles rip the tissues of the lungs as grit would damage a silk handkerchief. Both of these diseases have recently been classified together under the name of pneumoconiosis.

Our travel was to some abandoned workings in this smaller seam. We needed some sets of steel arches before we could rip our roof any farther, and they were in short supply from outside. In this old working were some good arches which were not doing much good in their present position, so we were to salvage them. After passing through two doorways the atmosphere became warm and very stuffy. Soon George was showing a face gleamed as if black-leaded. Wiping off the sweat with a shirtsleeve put a nice minstrely polish on his chubby features. He kept behind me, conscious of the stale, mouldy smell and the sense of unreality which always pervades mine workings when men are not there every day. Woodlice were busy on the timber, beetles and flies flew into our faces. Many old bits of wood were lying about the side, but when George sampled one with the intention of taking it home for firing it crumpled to pieces in his hand.

A long way down a solitary light showed us where Jim Lamey was keeping his guard over the pumps. They were spluttering away and the water gurgled outwards along the columns of pipes. Jim had his chalk marks on the side to show how fast the water was receding. His eyes were goggling at the need of sleep, and as I was an old friend our arrival was welcome. He insisted on our sitting on his damp seat while we talked mine gossip—'This place has fallen in, that place has run into a "jump up" of the coal'— and sampled the chewing of some Spanish liquorice root which he had to while away his time. It seemed exactly like chewing bark to me, but George appeared to like it and Jim's jaw was wagging busily.

'If a chap could smoke now,' Jim explained, 'it wouldn't be so bad. That is if he could afford enough tobacco.'

'Well, he hasn't taxed it any more,' I said, for it was the day after the Budget speech, 'nor is there any more on beer.'

'Get on,' argued Jim. 'They were ashamed to charge us any more for water.' Jim's job was lonely if not hard. Usually his only companions were the rats and beetles; 'or one of them bosses as comes a creeping round with their little night lights.' Hundreds of yards in that area were under water, and the pumps had to be kept going, else some might flow into the other parts. That mine water was coated with coal dust and coloured with iron mineral. It would ruin anything it wetted, and it stank—horribly. So did everything, even Jim, that came into near contact with it. It gave off a filthy, nauseating smell which made us think of sewage, and that reminder filled the galleries and revolted our noses for hours after we got again to the fresher air.

'I'll come along and give you a hand,' Jim suggested, as he hobbled along with us. 'It'll warm me up a bit. This hole 'ull rot a blinkin' bloke away, and it's a blasted good job as I leaves these here clothes in the baths.'

He was a permanently crippled miner doing this pumping job as light work. Jim was one of the few who could be accommodated out of the many who were seeking light employment. Farther down again we came to the steel arches with about a yard of the foul water along their sides. We collected rotten timber, filthy stones, and some old road sleepers. With these we made a pyramid on which to stand and reach up to the eleven-foot-high fastenings of the arches. There they were bolted together, and we had to unfasten those bolts. Each time we placed anything in the water we held our noses and waited for the poison to clear out of the scanty air current. It would have been a disaster if one of us had fallen into the horrible pool which stretched far away into the darkness—not so much because of the wetting, but because of the shunning which would have been our penance for a long time after. That unbolting is a dangerous job, as often the plates spring when the nuts are slackened. Fortunately they behaved well, and there came no cause to jump away swiftly to a smelly safety. Later, we fused a long chain on a leverage implement called a Sylvester, and dragged the big arches back from the water. Those big arches look graceful and not too cumbersome out in daylight and ample space—but in the narrow roadways they are like trying to force a battleship along a canal. We had to drag them for hundreds of yards, and it was a foot-by-foot job.

George, ever curious, wanted to know about their smaller seam. I showed him how the packing was more complete in this yard thickness, and why the cogging and timbering was so much easier. We bent to look under the roof where the coal had been.

'Want Snow White and the Seven Dwarfs to work this lot,' he stated.

'Snow White?' Jim chuckled. 'That's a good 'un. She won't be very white or sweet after she had bin in here.'

'Get away,' I argued, 'a yard seam is counted nice to work in. Many of the small seam men would think they were in clover if they had this amount of room to move about in. What would you say to working in eighteen inches, and p'raps water under you at that?'

'You won't get me in eighteen inches,' George objected. 'No sir. That's definite. No eighteen inches for me.'

'Wouldn't get you out of eighteen inches either,' I agreed, studying his waistline, although it is diminishing.

It is a fact that men do not like the smaller seams after working for a period in the bigger ones. The coal is usually not so easily got and must be dragged a longer distance. There is an advantage about being able to stand upright on your feet and being able to move quickly. There is also that freedom from stiff and aching knees, sometimes inflamed, and that continual knocking of elbows and heads which accompanies the small seams. Yet there is an early period when big height frightens a newcomer and makes him nervous of top which he cannot reach to sound with his mandrel.

You do not get coal falling out for you from the small seams. It has to be cut, and often that takes a deal of skill. In the big seams, when the place is crushing tons of coal will be squeezed free and ready for the filling. Of course, an attempt is made to graduate the cutting price to suit the seam. If we try to point an easy example it would be felt that if, say, three shillings a ton was a fair cutting price for the yard seam, then two shillings would be about the cutting value of a six-foot seam, and a shilling a ton for the nine-foot. That is only a bare instance, showing how the price might be varied where the coal is plentiful and easily got. As I have said, that can only be a rough guide, because stretches of the same seam vary so much in workability.

There comes a limit above which height is a handicap. Six feet of height is about right for mining, as then a man can fix supports against his roof and can test that roof for weaknesses. Above that it gradually gets harder to control, the pressure is more severe, and it becomes increasingly difficult to get material to fill up the spaces. Imagine something about eighteen feet high and think of dragging posts that weigh more than a hundredweight up there and fixing them in position. Stages must be built so that the timber can be worked up gradually. Judgement with strength is necessary, because there is rarely any spare room to twist that nine-foot or six-foot post about, and there is often an art in getting any light there at all. These high cavities are usually full of gas, which naturally works upwards. Sometimes a man is busy working when his head starts to swim, and the next thing he remembers is one of his mates asking plaintively:

'Why didn't you say as you were going to fall? Nearly dropped right on top of me, you did.'

Repairs in the smaller seams, again, are usually simpler. It is often a question of making more room by blasting the sides or the roof down. It is amazing what a difference six inches can make when setting up steel arches or timber; and I have seen men work hard all through a shift to gain that six inches.

The spur is tickling your ribs at the repairing work in similar fashion to the price inducement for piece workers. Here and there the management can find a tool who, for the sake of a few shillings more a week, and a pat on the back, will set an impossible pace for his fellows. Now I believe in a man doing a reasonable day's work, no matter what system controls his life, but even a machine must be run at a sensible pace. If you watch a builder, or an experienced farm labourer, at work, you will notice they keep a steady pace into which their thoughts and their actions merge as they go along until there shows to be no strain at all. That way they can work the next day, the day after, and for many years. His health, and his labour are usually all the assets that man has, and he must not overtax his reserves if he is going to have a long and useful life.

Imagine one of those workmen hurrying to his job, stripping to the waist on a cold morning and rushing at his work as if a fortune was buried beneath his feet and he had only five minutes to dig it out. I expect the lunatic van would be there very soon. Yet that is the method some bribe-mad miners employ. They are not many, nor do they last long, but by various means their depleted ranks are kept moving. If they can fill a ton of stone more than the next man, rip half a yard more, or stand more timber, they get their little extra pay and are most useful as man-work standards for the officials.

'Tom Slogger did more than you today. How was that?' or 'If Slogger filled ten trams, how was it you only filled nine?' This continual bantering affects some of the men. It spoils their tempers and strains their nerves. Always remember that undue haste may, and often does, mean death in the mine. Remember, too, that only one man gets this extra payment and all the others, down to the labourers, must do more work because of it. Hermit did a lot of this backhanding, and for some years it seemed he would stand the strain. Now he is cracking up and can feel it.

I remember one of this type who went to work in a place which others said ought to be made safe before the fall was clear. He rushed to clear the fall, not worrying about timber. An overman tried to get others there too. 'No. Let him get on with it,' suggested a seasoned old repairer. 'You see that the stretcher is ready. You shall hear him squealing before long.' A rather brutal forecast, but it very, very nearly came true.

I never forgot another comment made by a man who had always seemed an ideal type. He was quiet, decent in his life and always ready to help others. He had suffered from the bribe-takers, and once astounded me by saying: 'I've got to hate that sort of man. I think if he was under a fall and shouting for help, I'd walk up and kick him in the mouth.' Also amongst the piece workers, and the workers on the coal, it is far from usual for the best workman to get the highest wages. It often depends greatly on what pub he drinks in, or what chapel he attends, or whether he can be relied upon to do harm to his mates when required. In some local collieries a few men, a very few, worked as blacklegs during the trouble of 1926. In many cases they are still getting favoured treatment, and the miners watch it, with a bitter comment: 'The company have got a good memory. Pity we don't remember a bit better.'

The trouble does not end in work, because in our mining streets houses are so closely huddled that lives are almost intermingled, and people learn all about each other. In most of the company streets, two families live in the same house. When both men are doing the same work, yet one brings home more pay than the other—and often the wife flaunts it—then the friction starts in real earnest. I think a rigid flat rate, with an extra payment to suit each grade of work, is the wisest solution, and it should be possible for the men's union to discover what each man was paid. If some official had been especially kind to him there should be a way of finding out the reason.

The massed greyness of the average colliery street, when built by the colliery company, makes me shudder. It looks like a long stone wall with doors set in at regular intervals. I know that most of the houses inside are cosy, chiefly because the wives have been working every hour to make them so. With ample building material at hand ready to be quarried, with immense stretches of open land across which the fresh winds blow all around them, with cheap and plentiful electricity available, and often distributed by themselves, and with an immense supply of water always ready, why did those old builders commit these crimes against generations of humanity? And even if they were so cruelly blind, why have the following generations condoned such lack of decency?

Welcome Home, c. 1943–5, watercolour, 20 × 25

The condemned Lower Row at Blaencwm was still partly occupied. The house ironically marked 'Welcome Home' is boarded up but the one to the right appears through its open door to be still furnished.

Untitled, c. 1943–5,
pastel and pencil,
47 × 34

Isabel painted this miner from memory. He may perhaps have been waiting for a bus to work, or biding time having had to quit his job through injury or pneumoconiosis. It seems a portrait of bitterness and hopelessness; a man staring into his own future.

MINER'S DAY

Blaencwm, c. 1943–5, watercolour and pastel, 35 × 43

Condemned houses at Blaencwm have been left to fall in on themselves. This view looks through the collapsed end of Lower Terrace to Upper Terrace behind.

It is quite the accepted thing to be married ten or twelve years and still exist in one room up and one down, never having a home of your own. Politicians have promised without cease, even a king came to this area and promised something would be done, before he lost his crown, but promises do not stop hearts from going hard and sick with disappointment.[11] Until a few years ago a long row of dilapidated Army huts, bought second-hand after the war, sheltered—very inadequately—several mining families. They were tight alongside the main road for anyone to view

---

[11] Touring South Wales soon after he acceded to the throne in 1936, Edward VIII was quoted as saying of the unemployed in Dowlais: 'These works brought all these people here. Something should be done to get them at work again.' He abdicated later that year.

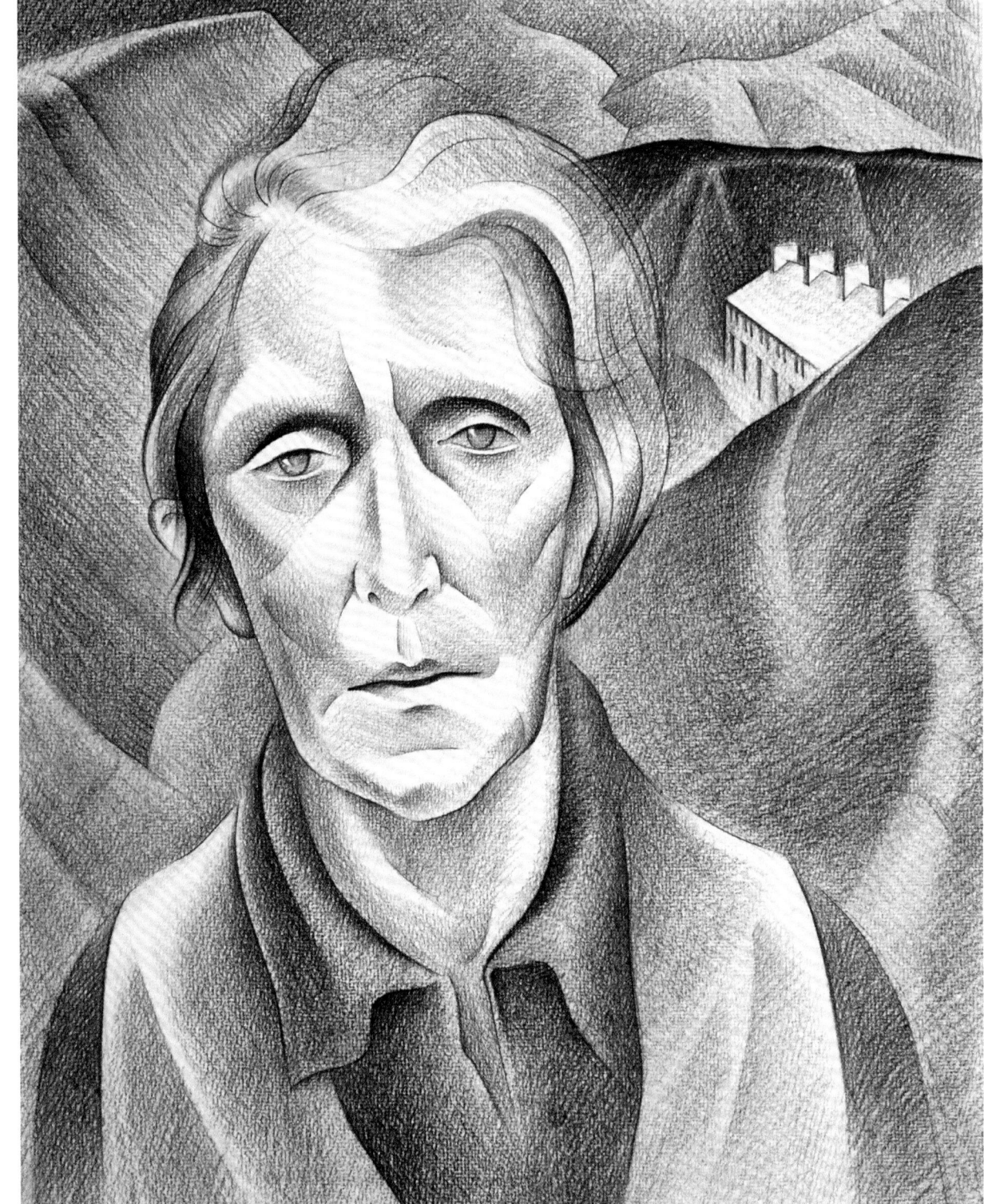

Mrs G., 1944,
lithograph, 38 × 30

This extraordinary portrait parallels Bert's comments about the lives of women in mining communities and the special economic challenges for families outside work. The lithograph was published in the illustrated essay 'Coal: The National Plague Spot' in 1946. The caption stated: 'Mrs G.'s husband is an ex-miner. Her sons and eldest grandson work in the pits. 53 shillings comes into her house every week. Rent is 12s 6d. Coal costs them 33s per ton, delivered. If Mr G. were still at work he would pay 6s a ton at the pithead plus cartage.' The setting is Blaencwm: Isabel recalled she sketched her near her home.

Penygraig, c. 1943–5, watercolour, 20 × 25

Rows of houses and their laundry poles are beset by tips that seem ready to engulf them. They seem to be the tips for Naval Colliery behind Tylacelyn Road at Penygraig. The sketch was reproduced to illustrate the essay 'Coal: The National Plague Spot' in 1946.

and the rents were high. They were horrible places, and because of the period they lasted we are dubious of any suggestion about modern temporary homes.

There are still scores of families living in wooden or corrugated bungalows here, but some of these do keep them neat and the surroundings attractive. Over in the next valley there are rows of dwellings worse than company houses. They are one-storey shacks built long ago when the collieries were started and still maintained. On most successful poultry farms the fowls are housed a great deal better.

One thing has eased the sores of these grey streets and made the house inside comfortable. When the rain—and we get more than our share because of the nearness of the

F. M., A Disabled Miner, 1944, pencil, 28 × 19

This naturalistic portrait was the frontispiece to the 1946 essay 'Coal: The National Plague Spot'. It echoes Bert's observations that men aged more quickly in the mines. 'F. M.' could be in his fifties but his face is drawn, his forehead marked by worry lines, and he stares into the uncertain future of a disabled worker. Isabel noted that she drew him in his greenhouse.

Blaencwm, c. 1943–5, watercolour, 34 × 28

In the essay 'Coal: The National Plague Spot' in 1946, this image was captioned: 'A row of eligible mining residences built to accommodate two families each. On the ground floor, two rooms and a kitchen. Upstairs, three or four bedrooms. Cold water tap (one) a feature. Bathing by tub in the kitchen. Outside lavatory.' It might be based on one of the newer streets in Blaencwm, such as Dilys Street, built around 1910 with larger houses and pavements outside.

sea and the mountains—washes down the slimed dust off the roof, and it clogs in the gutters and coats the flagstone which is outside each door as an advance note of the pavement which is to come, then the weary miner is sure there will be brightness and warmth inside; because his wife has carried on her never-ending battle, and has used the weapons that her grandmother used in the same house: soap, scrubbing-brush and water, and then brown paper over the oilcloth so that it shall not be soiled.

These mining women force an armistice from the dreariness, but at a cost to their looks and lives. Their hands roughen, their features become gaunt, and they become old when they should be mature and still attractive. I see no remedy in patching or altering the old streets in these places. The smudged slate must be wiped clean and modern homes, well spaced, and arranged so that housework is easier, must be built for the people. We are rightly concerned that our dairy stock should have good and healthy quarters. Can any reason deny that the workers also should have homes at least as sanitary as a cow's?

In the present crowded state the very young, and the very old, are an encumbrance, they have no place. The women wrap their shawls tight around the babies and shelter them with their love, but when they get older their living and sleeping becomes a problem. Youth can solace itself with a hopeful future, but what of the aged? They have outlived their usefulness, and with their children grown up have often lost the control of their own home. No quiet eventide for them. Their sons or their daughters marry without finding any other home, and soon love

Tylorstown, c. 1943–5, pencil, 21 × 28

A view of Tylorstown tip from behind Penrhys Road shows backyard sheds and laundry poles in the foreground and, across the Rhondda Fach valley, more terraces, a quarry and the incline up to the tips.

Tylorstown Tip, 1943, pencil, 20 × 24

Tylorstown tip was one of the largest in Wales and in its prominent hilltop position was visible from across the Bristol channel. This view is from the hill to the south-east and shows an older tip that underlay the conical construction with its tipper in action at the top. Both coal spoil and boiler ash were dumped and the tip sometimes burned, giving it the nickname 'Old Smokey'. A boy is carrying away a sack of gleaned coal and two distant figures are struggling down. The tip is now decapitated but still a landmark.

changes into boredom. The old folks cling to their ways which are not the ways of the younger family. Ofttimes the old man fails to work and must stay in the way by the fireside; then he becomes partly dependent on his married children for support and loses all independence. Sharing a common kitchen, and perhaps a bedroom, there is no privacy nor any quiet resting place.

I wonder, can wealthy, or even comfortably placed, folk realise the misery and hopelessness that this lack of houses can bring? I feel they cannot visualise it. If that problem were eased, I am sure our figures of suicides and murders would be greatly lessened. A small and peaceful abode—a Home of Rest—with a sufficient pension to make life possible: that should be the sure reward of every couple who have come to the evening of a life spent in hard work.

I noticed a sharp contrast whilst walking out this evening. At Cwmgrach the Welfare Commission have converted some waste ground into a little jewel of a park, with bowling green, pavilion, and tennis courts. Not far away, at the other end of a street, was a green meadow circled by the river. What an irresistible combination. A colliery company has started a slag tip on that green meadow and already the view along that street is being blocked by that increasing mountain of black waste from which the dust clouds on fine days and the slag flows away in flood times.

Nearby, across to the main road, is a lovely old villa which captured the heart of Southey when he passed this way.[12] A trivial argument led to a dispute which decided the landlord that this poet was no fit person to live in the locality and so Southey never got his desire. Then it must have been attractive; you can trace it as you can often recapture the beauty which once belonged to an ageing woman, but Industry is setting its mark close to it. That great tip, dominating the centre of the valley, is getting ever closer to the old house. They are building a memorial to you, Southey, these men who have probably never heard your name: a monument of ugliness and slovenliness that would be better used if packed back into the spaces of the mine from whence it came.

All day an automatic tipper climbs steadily upwards towards the top of the black pyramid. On the high peak the tipper seems to hesitate slightly as if ashamed, and then over goes another few tons and the tipper crawls downwards again. The dust cloud lifts and consorts with the wind.

Benjy admires that tipper and frequently tells me about it.

'Seems so cokeum, mun, just rolls 'em up and slings 'em over. Wish I could chuck out trams of muck like that. But there, s'pose as they'd give me all the more to chuck out, so it 'ud all amount to the same thing in the end.'

Which proves that Benjy has correctly analysed mining methods as far as machinery is concerned.

---

[12] In 1802, Robert Southey, Romantic poet and future Poet Laureate, sought a lease on Maes-gwyn House, half a mile from Cwmgwrach, and called it 'one of the loveliest spots in Britain'. The Williams family of Aberpergwm declined to let to a perceived radical. The house fell into ruin in the twentieth century.

CHAPTER SEVEN

Midday, and a warm sun toasting the mountain slopes. Gangs of youths climbing unwillingly towards their work and finding difficulty in keeping their feet near the dustiness of the pit incline; while a clear blue sky is reminding them of happy walks, and the land is abuzz with reminders that spring has come—but at the end of this destined journey the dark, internal dustiness of the pit is waiting to engulf them.

No wonder they call the afternoon shift the 'castor oil' shift. It cuts its workers off from everything. For them no evening walk, no cinema, no games and no social life in any form. This three-shift system has almost spoiled all the choirs, the bands and the dramatic societies which used to be so energetic in mining areas. What efficient practice can you have when a section of your members must be absent every second, or at least third, week?

Do the output figures justify this everlasting shift idea? I do not think so. The colliery has not time to cool, nor is there enough clear time for repairs and changes.

I know a colliery in England where the three-shift method was abandoned and the individual output increased; that output figure is still about double ours in South Wales. What extra coal is won by this middle-shift system is gained at a high cost in human happiness.

Have you ever watched someone of whom you were very fond walk away from you into darkness and into a danger which he could not realise? I have, and it was a time which tugged at my senses. It was the day when our boy started to work underground. It had come to that period in mining history—during the last three years—when folk seem determined that their sons shall stay out in the sunshine and fresh air. We did not wish our boy to start, but for some reason he suddenly insisted on going. I knew he was going to work on an engine, a safer and less laborious job than most, but still—something was dying in my inside when I watched his light disappearing after we had parted at the division on the main drift. I had gone with him as far as I could and had been sure that a reliable mate would take him the rest of the way. So I stood and watched that light go fainter along the dark tunnel. I wanted to hurry after him, call him back, take him again outside to safety, but somehow I did not move—and I guessed he would not be agreeable to any retreat. I wished I had touched his face or given him a last handshake—yet men, and boys, I suppose, are not good at emotional partings.

I thought, as do thousands more, why should our boy have to work in this unnatural place? Had he not the right to walk the land and see the sky every day?

Why should his body be battered and his skin blemished, or his lungs choked?

In one mining report I read, and always remembered, the accident figures to young miners under twenty over a period of ten years. It was over a quarter of a million. They are so fragile, these young lads; I wonder, do their bodies or their minds ever completely recover from a serious accident?

Possibly in my case it was not so much the bodily danger that I feared. I did not want his mind to become brutalised by the things he might see and hear underground. Coal is a hard mineral, and there are things to go with its winning which are as black and as hard as the coal. I did not worry about myself, from the danger point; I know all the tricks, and if I get caught by a fall, or something else, it will be something beyond the skill of a man to foresee, and as for the mental effect, both of language and treatment, I have hardened to that. No man now will make me go the way I distrust, or make me believe what I feel is false.

Yet a young and impressionable boy—but I need not have feared. He has by now chosen his companions, and they are quiet and thoughtful like himself. There are many such in our mines. And he came out to greet us at the end of the shift, a lot blacker, a lot wiser, but still active and whole. Many times since then he has travelled those roads under the earth, and our fears are eased into dormancy.

Yet still, one thing sears my soul. It brings my anger ablaze each time I remember it—and possibly I consider such things too frequently, but it would surely be better if many others considered them well. Shall we have an auction? Now, I know that most miners are dependent on industry's bidding. When coal is not needed they are an encumbrance, a glut on the human market, only of use in that their presence may help to frighten those who are still in employment, but when coal is needed they are treated as useful citizens, both by the masters and the country— until the rule of supply and demand again reverses the scale and the worker is pushed back undusted into his hiding place, or as the Welsh term it, *cwtch*.

Yet this auction I suggest is not in the use of man's hands and skill; it is in human lives. Take a look at Will Jones. He is intelligent and a decent citizen, who cares for his home and his wife. Tomorrow he may be brought home crushed. What is his value? No market haggler has ever developed his skill to a greater extent than the insurance assessor. Will is getting rather old, gone a lot in wind and eye. Shall we say two hundred pounds? What is that? Someone objecting over there. The Miners' Federation acting for the widow? Two hundred and seventy-five. Next, please. Name of Alf Slaughter. Not so respectable and well-intentioned as Will, but a deal younger, and might be fit to work for many more years. Three hundred and fifty? No? Has a wife totally dependent and you claim four hundred. My dear sir, that is the maximum even under the new order. Then back to a boy, starting life with high hopes and with the possibility of great things if we were a wise and far-seeing community. We will be generous, giving bare funeral expenses, something about twenty pounds. Not counted as a week's wages by some men, yet the value of a mine boy's body. And a judge will grant a thousand pounds and

W. M., A Young Miner, Rhondda, 1944, conté, 37 × 27

Isabel recorded this young miner with the initial 'W. M'. Bert wrote about his youthful colleagues with foreboding, knowing how quickly injury or pneumoconiosis could take away their futures.

Fred, Trealaw, 1943, conté, 37 × 27

Fred in Trealaw looks about fourteen, old enough to start work in the pit. Other than mining, the main alternatives for a boy of his age in the Rhondda was shop work as a delivery boy. Only if his family could afford to go without his wages would he be able to continue in school. Bert's family had a small farm to run but his son Peter still became a miner.

Young Miner, Rhondda, c. 1943–5, conté, 30 × 22

Isabel appears to have drawn this young miner after he came back from a shift, before he went into the baths. He is still black with coal dust and wearing his compressed-paper helmet and working clothes of scarf and jacket. The dust has caught in his stubble, but his lips and eyes are rubbed clean.

more damages to a man who is well rid of a faithless wife.

Last week I saw two young and rather alarmed faces sheltered by new and shiny safety helmets. The lads looked out of place, and they felt likewise. Their clothes were clean, a most unusual sight on that pit top, and their skin held the tint of warm suns; a contrast to the white washed-out colour of the other boys. They might have been mine inspectors going up—indeed, their garb reminded us of them—but seen at close quarters they were far too young and self-conscious. The experienced boys clustered behind them like schoolboys examining some newcomers. Probably they were interested in their outfit and the welcome given them by the officials—sensing that both would soon lose their freshness. Possibly, too, they were recalling the unheralded way in which they started their own working day. The newcomers were Bevin Boys arriving for their first day underground.

I feel this Bevin Boy experiment can have little effect on the output. If we cannot attract the local boys to mines, how can it be expected that boys from other areas will accept what they have left behind? The conditions of working and living will be so different.

If the boys see no future in mining what interest will they show in learning the craft? And if a boy does not want to learn and help he will be a hindrance and a danger to the man who has to work with him. Years ago I had a boy with me who was determined to get away from underground work. I do not remember ever working harder or being more worried in my life. He would get about into places where he was told not to go; and it was much easier to do the double work yourself than to undo the mess he made by sheer lack of interest. It was peacetime then and I understand he made a good policeman. Anyway I was glad and so were most of the surrounding workers to see him going.

I notice that some magistrates are placing heavy penalties on boys who have refused mining work after being selected. Several have been sent to prison. I cannot see what we shall gain by making criminals of them. It seemed that all of them were willing to serve their country in any place but the mines. Certainly prison will not make miners out of them or increase the output. It is far more likely to make them lifelong rebels. I wonder have the right inducements been offered?

In all the papers I have read the possibility of the Bevin Boys rising to high mining positions is stressed. I expect those boys, on seeing these enticements, would wonder what can be the matter with the local boys who have pit sense seeping into their minds all through their schooldays and have a heritage of mining knowledge in their own homes and yet would let such attractive positions go to others. As we have over two thousand mines in this country, and usually there is only one manager and one under-manager at each colliery, it is easy to see that the number of managerial jobs is limited, and sometimes the manager controls more than one colliery. No. The great majority of the three-quarters of a million and more workers in the mining industry must be prepared to do the more laborious jobs. Then why not instil the truth into the boys by telling them they will be learning a highly skilled craft which will help them to defeat nature and find danger with varied adventure right in the heart of a mountain? They can become part of a great union and through it learn how to take their part in local government, hospital administration and trade union work. They may be trained as delegates and take part in coalfield conferences. Many of the boys may have come from homes where democratic ideals are revered; they would not choose the severance that must surely come with the taking of an official post. Again, the boys would find that in mining work the manager gets little more respect than any ordinary labourer, and that in most cases it is the man that counts, not his label. Not all boys wish to become the blotting paper between the company directors and the workmen, and, with youth becoming increasingly intelligent, they may have decided for themselves that you cannot serve profits and humanity.

Anyway, the mirage that any mining boy may be carrying a manager's pencil, and nice efficient rubber, in his food box does not appear to be a very effective inducement. I am sure that the miners will give them a welcome and do what they can for their comfort. Some of the mining mothers may resent their coming and being rather pampered after their own children had been forced to leave the area, but I know that their bitterness was against the ones responsible, not the new boys.

There was no doubt about that forcing. I saw it even so late as just after the fall of France. Mines were idle and the men were called to the Exchanges to be told that work was available on airfields and in various outdoor jobs. Theoretically they signed their willingness to go, but when the dole is your only subsistence and you sense a threat even to that, there is not much freedom of choice. Besides, some of our leaders and others, who should have known, told us repeatedly and definitely that mining and the need for coal was finished, in these areas in any case.

I recall it was bitter weather and most of the men had worked for years in a warm temperature—under the earth's shelter. What it meant to some of them to be suddenly sent from home and out to work in the snow can be imagined. Usually the wife has learned how to ward away the little illnesses which beset her man. They missed that care, and many of them had unreasonable lodgings, so the toll from sickness was considerable.

Then summer came, the men were acclimatised, and they were able to work longer hours and increase their wages. Some prepared new homes; some had even moved their families. Then came a complete reversal; coal was urgently needed and all the men with mining experience must return to the mines. As a few months before they were not patriotic unless they went away from the mining districts, now they were ferreted out and told they were not patriotic unless they retraced their steps to the place they had been—and retraced them hurriedly.

Is it any wonder that these miners feel they are Aunt Sallies, to be shot at by any muddling official? Can we wonder that when they hear assurances of better prospects and secure jobs they tell you: 'Ah yes, we've heard something like that before. Many times.'

Yesterday I saw a picture of Mr Robert Foot, the new Chairman of the Mining Association, going into a mine.

It seemed that was the first time he had that pleasant experience, and the paper printed the fact as if it was an added qualification. He was talking to a mine boy, and this lad informed him that he hoped to become a colliery manager. A lad of modest ambition, yet possibly not very observant. With the evidence right before his eyes he should have known that the way to get a really worthwhile post in the mining world is to get the job first and go underground afterwards.

Of course it is possible for boys to become mine managers. It always has been. They may also become Members of Parliament or Field Marshals. But in each case the opportunities are strictly limited. There does not seem much wrong with the present rules which govern the qualifications of managers, but there does appear to be a recent tendency to foster the purely theoretical man and overlook the man who also has practical experience. I mean the years of experience as a working miner, not the period of maundering around as a student, which some count as practical experience.

Benjy speaks, and he is a connoisseur at this type of thing.

'Wish my old man 'ud made me an official, I do. Most of 'em nowadays becomes bosses because they couldn't do hard work themselves and, by damn, they is the worst sods to us workers.'

Of course the advent of Bevin Boys has brought its crop of stories. Usually Benjy has a new one every night. That Bevin Boy who was directed along a long heading at the end of which he would find his workmate. Long after, so Benjy says, they went to look for him and found him standing by a planked ventilation door which he should have opened and passed through.

'Been knocking here ever since,' he complained, 'and nobody come to open the door.'

'What d'you think of that?' asked Benjy.

'A well-behaved boy,' I answered. 'He had been taught good manners. He was in a strange situation and how could he know there was no one within a hundred yards of that door?'

Benjy still cannot think why any boy would not blunder past any doorway in the mine. His own boyhood, to whatever extent it existed, seems to have passed from his memory.

He was more successful with the second yarn, about a Bevin boy who was sent to a mate on company work. It might almost have been Benjy, and this mate took him a distance back along the heading, where he warned him to hurry along and let him know if he saw the light from an oil lamp coming, because that would be an official.

'Another where is he and what time is it bloke,' Steve commented. Benjy considered for a few seconds as if the words had recalled something in his mind, then went on:

'An' all at once he seed a light coming and just at that minute some repairer started boring one of them compressed-air drills right by and this youngster runs back to his butty, calling: "Look out. He'll be here in a minute. He's on a motor bike."'

Very much more will have to be done for the youth of both sexes before the mining areas will retain them against the call of more pleasant and enlightened areas. The Welfare has done much, I know, but it needs to move faster and extend its youth services. I have watched the youths who become interested in the Boys' Club Movement. Whether it is that a certain type of boy is drawn to that idea, or whether the boys develop that way after being attracted I am not quite sure, but I am confident that, on average, the Club boy is more decent in his behaviour and more intelligent in his outlook.

It is amazing what difference there can be in buildings created by the same fund for the same purpose. In one part of our valley we have Resolven Welfare Hall; at another part is Glynneath with its Hall. Resolven Hall, although fairly new and well equipped internally, is built alongside the main road, and has a grim, factory-like appearance. Glynneath Hall, equally new and built by the same fund for the same purpose, is placed well back in a nice park, and the grounds are well tended. The surroundings help to soothe a man's mind and to further the ideals of the scheme, while at Resolven they repel you.

Yet internally I feel that Resolven fulfils the ideals of the Commission much better. The large upstairs concert room is reserved for cultural occasions when visiting or local dramatic societies perform, or good class operas or musical comedies are given. Educational and documentary films are shown there. Again in the reading-room and library, although in this way Glynneath is also well served, there is a great keenness after books and a continual discussion about literature and its making.

My complaint about Glynneath is that the regular nightly film shows have driven all other interests away. Unless you want to see a picture it is better to keep away after early evening, for you will get neither room nor peace. I do not believe it was the ideal to compete with other cinemas when the welfare halls were built. If there was need for other cinemas they would soon have sprung up. What the halls were for was to supply a permanent place where other types of culture should be nurtured, and I believe all that was intended was that they should be self-supporting. I suppose it depends on the mentality and outlook of the committee.

This welfare levy of a penny a ton on coal raised, implemented by a payment from royalties, comes to nearly a million pounds a year: a surprising figure, showing what a vast tonnage must come out of the earth. It does seem to us in the industry that the scheme is slow-moving, and there are so many demands that large sums should not be allowed to accumulate. So far nearly seven million pounds, or about one-third of the levy to date, has been spent on pithead baths, and more than three hundred and sixty-five collieries are so equipped. It must be remembered that once these Welfare buildings are complete they are handed over to a joint committee of the management and the workers, who must keep them financially sound. We pay tenpence a week per man for the baths, but it is still a cheap and grand benefit. Canteens are another part of Welfare work, and for some time the local mine canteens were much shorter of food than the factory canteens around here. That problem is being solved. These canteens, too, must be self-supporting, but no large profits are intended. The institutes, parks, and playing fields section is another side of the Welfare Commission work, and they are getting ahead with them.

So the time must be coming when they will have to consider new outlets for the work, and of course something must be done about old age and about holiday homes. We already have some convalescent homes. Youth must then be considered in far more ambitious ways than previously. I feel that an extra penny a ton levy devoted solely to youth would not be a dear substitute for the fresh air and sunshine of which their work robs them. Nor need it be laid on the consumer, for I am sure there is a big enough gap between the cost of coal when it leaves the pit and that when it is delivered for the burner. Often there is more than a three pounds per ton gap, and I think a penny or two could be spared for the young folks. I would like to see a College on the lines of Impington in every mining village.[13]

One of my mates and myself started to put the world right verbally last night. It is usually a dry job, and as we had wisely started near a public house, we strolled in to loosen the argumentative valves. Wednesday would be a slack night, but although we did start a debate we found no room for sitting. We bore with the continual toe treading, beer spilling, elbow digging, followed by voluble apologies for the duration of two half-pints each, then wandered back into the street convinced that we would have to start on the public houses.

---

[13] The visionary Impington Village College in Cambridgeshire, opened in 1939, was devised by Henry Morris to bring together a school, community facilities and adult education in buildings designed by Walter Gropius and Maxwell Fry.

I still feel there is something in that decision. I do not like sawdust-smirched drinking dens. I want room and sometimes quiet where I can talk to a friend or read a book or listen to intelligent talk, and I like to sip my rare drinks slowly and appreciatively. I have not yet allowed my throat to become a sewer, or my mind.

Yesterday I saw a Bevin Boy groping about for his clothes in the clean clothes section. The place was well lighted, yet he was feeling for them. I went to help him and he thanked me, courteously and in a cultured voice. Apparently he had mislaid his glasses, for another lad brought them across to him soon after. I noticed how powerful those lenses were, which confirmed my fear that the boy's eyes were very weak. What concerned me was the type of medical examination that lad must have had before he was passed for the mines. With such eyesight he must have been a danger to himself and to others.

Obviously spectacles are no solution for work underground. The steaming atmosphere and continual dust would make them of little use.

One day I noticed three young lads in the canteen. They were intense over some sketching. As our boy was there I had a good look at this so interesting thing. 'This here is the throttle, but it's a wheel control, not a lever like this is on our engines; and the brake have got to be tightened with a wheel. And this here is the clutch and that works in a different way.' One of them had been watching an electric engine at work, and was explaining the difference to his friends, who had always driven compressed-air engines.

Why do they not start simple engineering classes in

the mining villages and give the boys privilege to attend? Usually they are scared by the idea of attending a technical school straight away, but with interest and confidence once aroused they would probably go on farther. There may be a chance for a couple of thousand more trained mechanics in the mines as machinery gets increasingly used, but I cannot imagine any greater increase being necessary. The underground fitters are paid at rates which vary with the skill they are deemed to possess, and the most skilled get about the rate of repairers.

Apart from other machinery the haulage engines in a pit are a varied and mixed assortment. Their drivers talk about 'a pair of eights,' or 'a pair of tens,' all descriptive of the strength of their charges, and promotion from the smaller to the bigger is a matter for careful watching and strict seniority. We use little 'Jimmy Wildes,' which can be tucked under any overhanging piece of side and have about the strength of two horses,[14] on up through the various sizes until you come to the lovely triple-drummed electric engines in their steel-ribbed and sheeted underground palaces sometimes twenty feet high and equally wide with a length of eighty feet. Besides these there are the slick-moving engines at the pithead which are on view each day. Thinking of engine drivers, we have just had a claim for a youngster disabled by dust who has only worked as an underground engine driver. Obviously the steel rope continually winding in and out carries with it some of the dust from the floor of the mine roadway. He must have been inhaling that, and it has affected his lung in a short time. We can think of no other explanation. Benjy jumped in on an argument about engines and horsepower to extol a new lamp he had fitted in his home. 'That's a good 'un. Aye by damn. Eighty horsepower it is.' It did seem a marvellous lamp, but Benjy's radiance was dimmed somewhat when we discovered he meant candlepower instead of horsepower. 'They're all the same to me whatever,' Benjy insisted loftily, 'for by Jawch, it do give a good light.'

Then Crush Williams, to whom nothing smaller than a thousand per cent exaggeration is worth considering, knocked the confidence right out of Johnny Rees in an argument about engines.

'You gassing about engine driving,' commented Crush, 'and don't know no more about it than me granny who have been dead for twenty years! I useter drive one in north Wales as had thirty-five clocks that I got to watch.'

Pithead Gear, 1944, lithograph, 7.5 × 7

An illustration of a headframe for the cover of *Miners Day*. Hundreds still stood in South Wales; this design matches those at Naval Colliery, Penygraig.

---

[14] Jimmy Wilde, height 5ft 2in, was a miner in Tylorstown, Rhondda, before becoming a professional boxer. Nicknamed, 'the Mighty Atom', in 1916 he became the first official world flyweight champion.

That finished Johnny. He was moody for hours. Later in that shift I went past his engine house. He was standing by the throttle looking into the darkness, and I heard him repeating to himself: 'Thirty-five clocks.' Obviously he was recreating that engine with row after row of clocks with Crush's mind and sight flashing like quicksilver amongst them in an efficiency that Johnny could never hope to achieve.

I told Crush that Johnny was worrying.

'It's right enough I did have one,' he insisted.

'Get away,' I argued, 'you were drunk and saw every one five times over.'

'Never been drunk in me blasted life,' he claimed, 'anyway not since last Saturday. And let me tell you'—some trams off the road had upset his non-angelic temper—'as I've bin to Africa, I've bin to America, I've bin to France, I've bin to Gallipolo, and I've bin most all over this blinkin' world, but this is the wust blasted hole I've ever bin in so far. It is that.' That way we are informed by Crush whenever the least thing goes wrong. He has been to the places he mentions, but as he probably gave each its turn of being 'the worst blasted 'ole' we are not greatly concerned about his judgment.

I watch the mine boys every morning, dragging their boots over the wooden sleepers, or through the coal dust and usually flexing their arm so that the lamp shall not hit the ground. Their small, white faces are impish as they grin from under a cap flopped to the side of their head and meeting the muffler that is always tied, for some reason, alongside their left ear. The jackets and trousers are usually torn, and always much too large. Continual kneeling has worn holes in their trousers and the knees gleam through, and also the small coal must glide through while they work. A tea jack bulges from one jacket pocket and a food tin from the other. Threadings of detonator wire hold the clothes from falling apart—they mostly throw them in the dirty locker and forget about torn pit clothes until the next shift.[15] After the shower bath they emerge with their hair expertly slicked, but shadows of coal dust around their eyes. A pie and a cup of tea in the canteen, then they are away by bus or train to their homes.

In many ways things are easier for them than twenty years ago, but so they are in all industries, and improvements will have to come much more swiftly if mining is not to become an industry of the middle-aged without any replacements. Some of the colliers, too, must share in the blame. Many of them, whilst demanding fair treatment for themselves, have still been eager to make a profit from the labour of the boy who works with them. A few have forgotten that it is a tender human plant which works with them, and imagined it was a machine.

---

[15] Pithead baths provided from the Miners' Welfare Fund had clean and dirty sides with showers between. Lockers on each side were ventilated with hot air to dry wet clothes. Miners put their working clothes in the dirty lockers, showered, then dressed from the clean lockers to go home.

CHAPTER EIGHT

'Look right over there,' called Peter, our son. 'Up in the sky.' That sky had been shuddering for quite a while under the continual passing of aeroplanes, and as our home is high up they appear to be coming straight towards us. We looked up at his suggestion and saw a sight which checked our breathing. The half-darkened sky was starred with moving lights, red, blue and seeming yellow, which came towards and over us to the accompaniment of that unceasing flying roar. It took us some seconds to realise that they were planes, a sky full of them, flying with all the lights on. Before the roar from one group had faded, the sound of the next was spreading into our hearing. We had seen planes in plenty every day, but nothing like this massed and continual flight with lights on.

One thought exulted us, thrilled us, and yet saddened us. We were sure it was the beginning of the invasion of Europe, and our saddening was because of those young lives moving above us who might never come back. The inside of the earth was electric that night, and in our spare periods we talked of little else. Morning came and we hurried out. We studied the faces and asked questions of the day men when they came up to meet us, and we rushed down to hear the first radio news. What we expected was not mentioned. For a short while we surmised it might be kept secret for some reason, but finally we had to decide it could have been no more than a rehearsal.

That evening a little boy, about five years old, was missing. He had done some mischief which had brought a threat of smacking from his mother. The boy had run off and disappeared. When dusk came with no sign of his return we began a search. There are many mining fissures and old pits about these mountains, and it was a grisly business probing about these. We had one bad moment or so when we saw something wedged down a fissure. Very glad we were when examination showed it was the body of a sheep. After many of us had hunted for a couple of hours we returned to his home to say we could not trace him. His mother, standing by the back door, screaming hysterically when she heard that, and suddenly there came a whimpering answer from quite near—in the tiny coalhouse. The boy was there. We had looked in without seeing him, and no wonder. He had placed an old sack between two large lumps of coal, had put another sack over himself as a blanket, and had completed the covering by drawing a sheet of zinc down over his head. What had inspired his infant mind to that method of simulating an underground fall as cover? Some talk he had heard in the home, probably. On the mountain peak high above us is a

Children in the Streets, January, 1944, pencil, 26 × 15

Isabel drew feet and legs (page 40) in January 1944, possibly at Blaencwm where she drew several children's portraits. A girl holds a toddler without shoes. She is wearing slippers and a thin coat. Her bowed legs and thickened ankles are evidence of rickets caused by deficient diet in the Depression. There is a pavement up to the wall of the house; a bucket on the curb suggests it has just been washed.

stone cairn built to show where they found the body of an equally young boy who, to get to that lonely point, must have struggled several miles from his home at Aberdare. It is a grim place, looking across the lesser mountains, and alongside a large lake which has a sinister reputation.

We have various ways of getting news underground. Often the lamp men hear something which they tell the rope rider who goes underground from the surface; he tells the rider who takes the empty tub off his rope. That one tells someone working along one of the headings, and he tells it to anyone he may see. It is not a reliable service, any more than some of the anonymous messages which come along the underground telephone are. If it starts fairly accurately it may not be so badly distorted when we get it, but if it starts in a jumble and gets mixed on the way, then we get some surprising information. Also frequently we have a messenger who sets out to mislead us. If a big fight is on, or a horse race, scores of men sweating underground are asking or wondering who won that fight or that race which happened whilst they were away from information. Messages seep in from those who are similarly interested—but are they reliable? If our fancy has won, or is stated to have won, we pretend not to believe the message. If our fancy is stated to have lost, we pretend a belief which we, secretly, hope will be upset when we arrive outside. We get our cuttings expert with his pieces pinned on roof posts, sometimes an attractive lady, but more often a sobering one like that quotation from Major Lloyd George's[16] statement concerning mine accidents—that in the last twenty-one weeks two hundred and fifty-six miners had been killed and one thousand and fifty seriously injured. Seriously injured—they must have had it pretty bad to be classed that way, and very surely the lighter casualties must be about ten times that number.

Figures which alarmed me much more were those handed to me by a mate who had been to an area meeting

---

[16] Gwilym Lloyd George, son of the former prime minister, was Minister of Fuel and Power in the wartime government.

of west Wales miners. These showed that from January 1 to April 30, 1944, in this one area of the coalfield (west Wales), two thousand four hundred and forty men had been stopped from working underground because their doctors suspected some dust disease. The number certified for compensation was six hundred and sixty-three. One thousand one hundred and fifty-two were awaiting a decision. In one small mining area—it is a ghastly affair. Obviously, if they cannot soon find some sure preventative for this disease, humanity—and the shortage of miners—must compel them to close down.

Not all the coalfields are so badly affected. It is the anthracite area where the coal is so hard, with the result that the coal dust is sharp and hard also. I heard, or rather joined in with, two men yesterday who were interested in this matter. One asserted that by using what we call curling boxes—like large weighing scoops—only one-sixth of the dust is created as compared with shovel filling. Not all the men like using, or letting their boys use, these boxes. In a high seam this attitude is understandable because the coal slips roll out for a long way, and may come over when the boy is scraping coal into the box, but in a smaller seam the risks are not so bad.

According to the explosive and emphatic statement of our house coal committeeman, who weighs at least sixteen stone, there are a couple of hundred men, householders, at this colliery who are weeks behind with their monthly allowance of house coal. Then men browbeat him because they cannot buy or borrow coal anywhere else, and he passes on the bad humour to us. 'What are you going to do about it, hey?' Not the hardy annual, but the hardy monthly, or often hardy weekly. Almost continually, for several years, I have heard that complaint brought to our committee meeting, and with it, lately, that consistent complaint about excessive dust.

Pipes have arrived for carrying water inside, but they have not come in large enough quantities, and the complaint is that those on the pit are not being used to the best advantage. One complication is that, after being indifferent for a long while, all the miners in the area have become alarmed by this spread of anthracosis and are insisting on water-carrying pipes being brought to their colliery so that the supply is not coming through fast enough.

We are negotiating about a cutting price for the coal in a new part which is being worked. It is a small seam, somewhere in the thickness of a yard. The suggested payment is five shillings a ton all in. That would mean that all other work would have to be without any payment, as the five shillings is to include all costs. Posting and cogging would have to be done to keep the places safe for filling, and the cutting face would have to be kept clean and in line. That five shillings would include this work. It appears a fair price, and higher by far than in most of this coalfield. The discussion is only beginning as yet, and no settlement has been reached.

I saw a pay docket today with a visitor from another colliery area. Forty-one tons of coal had been filled for payment of five pounds two shillings. He showed an instance of another collier filling five tons less, yet getting two pounds more pay. How can that happen? The old bogey of allowances, of course.

If a man is doing work which frequently takes him from outside a colliery to the inside, is he a surface workman or an underground worker? Say he spends half his working time underground and the other half out, must he be paid as an underground man or at the surface rate? For the Porter award has made a difference of ten shillings. Again, it is claimed that the increased percentage award does not completely cover the extra some colliers must pay to the boys between seventeen and eighteen. It seems that mining payment gets more complicated with each award. I wonder, has the person outside our industry any idea how complicated it has become? It may be a good idea to give a couple of examples.

Pay Sheets, 1945

Two wartime pay dockets were typeset in the original edition of *Miners Day* to show the large number of potential deductions from wages. One docket was Bert's own for 3 April 1944, which he interpreted in his text. The other was a blank form for another colliery, showing even more categories of deduction.

Just a few explanations. It will be seen that there is more than six days on this docket. Six are paid at the highest day rate, and the extra shift is paid at a slightly lower rate because it was not for the same type of work. The Porter award is three shillings, which would have made the pay up to five pounds if that seventh shift had not been included. Coal has been bought and its haulage paid; that would be a monthly charge. The doctor gets twopence in the pound earned. The B.I. after 'Hospital' means Blind Institute. In this case no rent or bus charges have been paid, but the Income Tax man has taken a larger bite than usual. The six shifts at seven-and-sixpence and the extra one at seven-and-threepence have been counted together to make two pounds twelve shillings and threepence. Below that sum is the percentage figure ruling at this time; below that again are the attendance bonus and the war awards which would, had a six-day week been worked, bring the wage up to three shillings short of the four pounds eighteen and six minimum of the Porter award. Eighteenpence weekly is deducted from the five pounds, because I get my allowance of coal at a reduced rate.

Surely it is full time that some simpler method was adopted and all percentages ended. Why not say the man has worked five or six or seven shifts at so much a shift? If it was six shifts at sixteen and eight a shift, for example, how much easier for all concerned to understand!

I have been reading a book published exactly one hundred years ago. It deals with mining accidents and ventilation. The author wrote so simply and directly that his book does not in any way seem old-fashioned, and if it had not been for the dates and the fading of the very strong cover I would not have imagined it was so very old. 'Say a quarter of a century ago,' I would have said, 'and not much more.' The author was Joshua Richardson, F.G.S., and in his time he must have been counted a Progressive.[17] His heart and his instincts were in the right place, for most of his conclusions are scathing. He tells of a boy of nine driving a winding engine. He was frightened by a mouse and overwound, causing several fatalities. Poor frightened little boy! Amongst several instances is one where, in spite of warnings, the coal was worked so far out to sea that when water finally burst its way into some mine, timber was afterwards seen floating out through the aperture. The workmen were drowned and were apparently never recovered. He says that mining is 'a life of great danger both for man and child.' He records that very rarely was any inquiry made into the causes of a fatal accident. He says that 'there is a strong repugnance prevalent amongst the proprietors and directors of mines to all Government interference with the management.' I think his book helped to get some reforms; anyway he demanded them, and I cannot imagine him being very popular with his class after that book.

He explains the ventilation at considerable length and tells us of the lighted fires which were used to draw the air through the workings. I broke through into some old workings when repairing and widening an air shaft.

---

[17] The engineer Joshua Richardson from County Durham settled in Neath, where his book, *On the Prevention of Accidents in Mines*, was published in 1848.

Right at the bottom of an old pit I came to an old grate with cinders in it, coal at its side, and a bundle of sticks lying convenient. Alone in the earth there, half a mile surely from any other man, I sat and looked at the last act done by that air watcher who has been there, placing all ready for his next shift—almost eighty years before, as I found out afterwards.

Yet I, too, have seen fires lighted during my mining in this last quarter of a century, to draw air through workings. He talks of men having to work with candles, having not enough air to keep those candles alight, and collapsing from exhaustion. In spite of his appeals, that was happening seventy-five years afterwards, because I saw it all in the same way as described in that book.

We no longer allow children to drive engines or go underground so young. The examination into accidents is keen, and they have failed in their 'repugnance to Government interference,' but with all their research and sincere interest in our safety they have failed to lower that terrible accident and disease rate. Now we have giant fans to suck the air through the workings, yet the general scheme of ventilation appears much the same, and the ways of working by hand.

We get compensation—inadequate, I know—and have some control over our payment and our working conditions; but we have had to battle for each improvement. This book speaks of chest troubles and mentions them as bronchitis and other complaints. Obviously they have been existent for years, but more mildly than at present. The added demand for anthracite and the extension of machine mining must be responsible for our present state. The author complains that five years have passed, and nothing tangible has been done to alter mining conditions. Five years! What an impatient man! He would not be much use in the mining world in the present day. We are still waiting for many of the findings of the Sankey Commission to be implemented.[18]

Gardening time has arrived again. We are usually rather late in these areas. One reason is that the ground is so bleak and growth would not start until long after other areas. A more definite reason is that the average miner fears to stand on ground outside which is damp—or in chill winds. His work makes him more tender and susceptible to colds than an outside worker, so he waits for the ground to get warmed through before he arrives all ready to 'make this here spade talk.' We have some good gardeners and others. I am against getting too extreme in anything. The man who talks garden for all his waking hours is quite as boring as the man who has a political cure for bad weather, bad health, bad payment and bad nature. I like to see a man keen, but also to spread his interests.

These extra keen gardeners cause a lot of trouble. They throw the salmon tins and milk tins back to their proper owners, and that usually starts a row. They do not believe that radish and lettuce seeds are good for neighbours' fowls, and even the local children get no welcome when their ball or top arrives with a crash on the cucumber frame.

---

[18] A Royal Commission on the coal industry after the First World War chaired by Sir John Sankey made recommendations in 1919 about conditions and considered options for nationalisation.

Allotment, Rhondda,
1943, watercolour, 28 x 37

Isabel's watercolour technique was deft and confident, even when she worked in situ. Every wash and mark here is perfectly judged, from the figure made from just three or four brushstrokes to the grey sky and the granular mass of the tips. The allotments were in an unpromising location, on a moorland slope 750 feet above sea level, west of Penygraig, with power poles marching through them. The outline still exists of the abandoned plots behind Tonypandy Community College. The tips, now cleared, are shown again on pages 19 and 94.

I have seen some quite good ways of dodging gardening without having the shame of seeing weeds grow unchecked. One long narrow garden with a narrow path was altered into a long garden with a very wide path, and that answered very well. Two or three strong and quick-growing flowering shrubs will soon fill the average garden, and then there is obviously no reason for digging. A deal of skill is being used locally in covering the front gardens with cemented designs. It seems an excellent alibi. Then leave just a little lawn to hold the deck chair, and with a kiss of the sun the Saturday evening is complete. Over the fence that keen gardener—poor old duffer—is slogging away at digging and watering.

Yet somehow there surely comes a time when that gardener starts getting things—new potatoes, green peas, salads—and so many more nice things. Then the deck chair is empty, while its owner leans against the fence and explains how miserable life is if you cannot have an occasional meal of something fresh from the garden like potatoes or peas, and so much healthier when home grown. If it was not for those great bushes, or that cement, he would have tried to grow… No good! 'The mean blighter has gone inside with his load, and never even offered us a meal of potatoes, and I suppose he will be spouting about Socialism one of these days again.' Seriously now, without offering any political opinion, I have begun to think that many people's idea of Socialism is that of the man who leans on the garden fence. He thinks the work a waste of time, but he definitely wants a share of the results if they are good.

There is one man to whom it is not wise to mention the pleasures of gardening. That man is Crush. Saturday afternoon came, with his wife away on a short visit. He decided to become a reformed character. No public house that afternoon, and he, so he claims, cleared the house up 'like a new pin.' Looking for new worlds to conquer and something to take his mind off his usual habits, he suddenly recalled that other people dug their garden. Finding an old miner's shovel he got to work—and Crush has some power and energy when he feels the desire to use it. Late evening, with the patch looking very neat and freshly turned, found Crush, still with some energy left. In the pantry he found part of a sack of potatoes, opened out the rows and planted the lot. 'And by damn, we would have some spuds the way I shoved 'em in.'

Next morning his wife was delighted with all the changes. To do her share she wanted to cook an extra nice dinner, but it was awkward because extensive hunting failed to find her store of potatoes. Yet they had to have dinner, and to make that possible Crush had to retrieve them from the garden where he had planted them so carefully.

'Aye, on Sunday morning, and all them togged-up articles as wus a-going to chapel staring at me as they was going by. If one of 'em had said a word to me about breaking of the Sabbath I 'ud a rubbed his nose in the potato row. I w'ud that.'

So one recruit is lost to horticulture, and Crush unloads pints for a change and glowers at anyone who looks as if he might enjoy gardening.

CHAPTER NINE

'If I had me way,' Crush asserted, 'we'd all go to jail instead of paying it. That's what we oughter do.'

'Wouldn't be no worse than this blasted hole,' Benjy agreed morosely. 'Wouldn't have to chuck muck all the time in a place like a blinkin' oven.'

'Might have to crack some stones,' Steve sounded hopeful. 'And it 'ud be good practice for old George. Cost 'em something to buy a suit with broad arrows on for him, though.'

They had been given bad news and were grouped together … dismayed and disgruntled. Weeks ago there had been a short strike of less than a week. As we understood it a threat had been made to stop the issue of cheap coal to men suffering from industrial diseases who were idle at home. Failing to get an undertaking that this would not be done, the men had gone home and remained there until the point had been cleared up temporarily by the men having their coal. Summonses had been taken out, and there was no doubt that a breach of contract had been committed. At first eleven pounds a man damages was claimed. It passes my comprehension how a colliery which is stated to be losing money on its normal working can lose a sum, when they do not work for five days, equal to eleven pounds damages per man out of over eight hundred men and boys. The boys would pay only their proportion of this, but the boys or youth were in a minority. Later this amount was reduced to three pounds a man and half that amount for boys. That we had to pay. As this would come unfairly on the men who only drew a regular day wage, we are trying to get a reduction for them.

We have avoided this type of stoppage before at this colliery and I cannot recall one happening previously over many years. I suppose the intention is to teach the men a lesson, and it will. That lesson will not be to abandon striking, but that in future any talk of co-operation or friendly discussion is so much waste wind. Was it worthwhile to be so vindictive?

I was glad that George was not in work when I went through the faces of the small seams. You do not learn how to compress your body into a small space at his age, and he would have a deal of that compressing to do. Besides, there is a terrifying sense of being crushed and smothered until you get used to the roof being tight against your back and your face not far off the floor of the seam. You have to wriggle around posts, scrape past coal, and be careful not to jar the bones of your knees too hard and often on the solid stone floor.

What a difference one inch makes! You scrape a channel

through the small coal and crawl along without rubbing the skin off your back. You study the roof and the sides and conjecture what would happen if that roof suddenly lowered a couple of inches or the sides were forced in— neither of which is impossible. Nor is there any short crawl about those conveyor faces either, since they go for a hundred yards or more. That is a long way when a man is plunging desperately, like a swimmer with a crawl stroke, to get along swiftly from something which is crashing behind him. There are the elements of a nightmare in that. You want to travel fast, and your fear insists that you rush out to greater height, but all you can manage is a hand over hand and knee after knee crawl.

The thudding growl of the coal-cutter picks smashing into the coal seam, and, undercutting for about four feet as the whirring turbine draws it along at a yard a minute, reminds me of the old time when I used to drive such a machine. I am not opposed to machine mining. Instead I welcome it when it eases the work without adding to the risks. Cutting by hand under such a seam was deadly monotonous work which used to crack up our hands and our elbows. Assuming that we get a ton of coal from each cubic yard, then in this yard seam they would be cutting a ton a minute. It would take a good collier all day to hand-cut the amount this machine does in four minutes. Let them cut, I say, but let it be done steadily and with every regard to safety or health. Things are getting better even with machine cutting, as they now have a fixture like a tank to spray water over the cutting chain and the dust is prevented from rising. I could hardly believe this machine was cutting, so different was the result from the choking clouds of dust to which we had been accustomed. I recall a time at the end of our shift when we used to brush the dust off our clothes and our faces with a sweeping brush. Certainly this spraying is a big improvement over the other method of boring long holes into the coal and filling them with jets of water which are plugged and left. Slowly and silently the water works along between the slips where the dust collects. It loosens the coal and damps the dust, making the coal face a much more pleasant place in which to work. I have crawled along these coal faces and, except for his voice raised to a shout to surmount the clatter, I could not recognise a close friend because of the dust which surrounded him. When this water treatment is used the coal length is a different place, a place where one can see for forty or fifty yards and discern the lighted lamps easily —nor is there need for that continual coughing which used to be the accompaniment of the shovel clanging and the tin pan rattling of the conveyors. Those old-type trough conveyors were very noisy while working, and such noise is a danger underground where a man wants to hear the first warning of cracking roof. Also they were cumbersome when unbolted ready for daily move forward to a new stretch. You could not bend them, so the posts had to come out; a terrifying, risky business sometimes. Even if no fall happened then, the supports which had become solid under that roof were withdrawn and the fresh ones were not so reliable. Now I see a belting in use which is far less noisy and more easily shifted. They just pull out the steel locking pins and roll the belting up in lengths like stair carpets. Then the roller cradles are moved across one yard and the black carpet unwound over the cradles, the steel pins are reinserted, the conveyor engine is drawn across and the tension posts set at their angle, and all is ready for another stretch of coal to be conveyed into the trams at the end. Chocks have been placed in the cut under the coal to stop pressure clogging the cutting chain of the machine. These are withdrawn when coal-filling is to start and the coal comes down with more or less difficulty. It all depends on the skill, sometimes the luck, of the collier. With the coal down the assistants or the boys turn it just behind them on to that moving belt which takes it along like a black river.

Colliers, I said, but the purists in our job insist that these men are not colliers, as the machine does the actual coal-cutting, and not much skill is required for the work which is left. It is correct to say that the cutting was the hardest and most skilled work and that many of the conveyor colliers would not show up very well if confronted with a stiff coal face and an empty tram—nor do the boys learn much of their craft unless they have training at other mining work than machine mining. It reminds me of a time when I used to walk a great deal and enjoy it—then I got a bicycle. Travelling was so much easier and swifter that I lost my interest in walking and never did walk so well afterwards. In mining it is somewhat similar, as machine coal is so easily got that the miner cannot endure the long hours of hand-cutting without any great amount of coal coming as a reward.

Yet there comes a period or a place in most collieries where hand-cutting must be done, and even now the managements cannot find younger men who will do it efficiently. Machine mining has undoubtedly helped greatly in our coal production, and in easing the most laborious work. It will be used more and more in the future, and if we determine to handle it reasonably it can benefit both men and management. By 'reasonably' I mean setting out a plan of work that can be surely and comfortably completed in each shift—and abandoning the idea of rushing cuts and clearances for a spasmodic record. A regular amount on each day, and all safety needs carefully guarded.

I saw machines used sensibly during my weeks in the Leicestershire coalfield, for they were content with a cycle of operations which could be maintained without continual rush. During the last six years their output had not varied one hundred tons per shift, and on the man average they doubled that of our area, although their seams, in my opinion, were more difficult to handle. I saw something else there—trust and friendship between management and men, and a spirit which is different to our alarmed watchfulness of each side in the industry. Possibly one reason was that coal was not the only, or even the dominant, industry. Our coal seems to instil hardness into the men who control it, or try to control it, and where men see nothing else, or talk nothing else, but coal, then it seems that the gentler natures soon fade out. South Wales has so much bitter history to forget that generations will have to pass into the unknown before a better feeling controls men's actions. It will need new men with pleasanter backgrounds.

I worked for a while in an eighteen-inch seam. Eighteen inches of height in which to move and work! To measure the length of a post we used to press our elbow on the floor as we were lying down, clench our fist, and the distance from the elbow to the fist was the height of a post. We sawed the mandrel handles short and took out the handles of the shovels, replacing these with a little crutch on the end of the steel blade. We had no power to lift a stone or a lump of coal. Instead, at full length we pushed it before us and got power from our legs straightening against some solid object. Only for the quarter of an hour of food time were we able to stand up, or even kneel. Crawling blindly through the acrid smoke of the continual blastings we banged our heads against posts, stones, and sometimes coal before we knew what was in front of us. Often the same smoke stayed unmoving for hours. We used to stand outside on the roadways before our start and wonder how we were going to squeeze ourselves under that small space for another shift. A small pocket of gas caught once, and the blue flame danced up along the coal towards me, like a living and very evil spirit. I had three pills of dynamite and a detonator in my hand ready to charge a shot hole. In such a small space rapid retreat was impossible, so I threw the explosives forward and lay with my hand shielded by my arms and my face pushed into the small coal. After some minutes, which I expected to herald my passage to eternity, I realised that nothing had happened. I uncovered, to find that by some miracle the blue flame had passed the dynamite and left the place unharmed. I had a ball of clay ready for ramming the hole at my side. This clay usually feels cold and clammy. That day it felt warm and comforting when I could check the trembling of my hand long enough to pick it up.

We used to massage each other with oil, like footballers before a stiff game, at the beginning of each shift and every night our wives used to put ointment on our grazed backs and raw knees.

'The more coal you fill, the more we will pay you,' the manager of that seam told me one morning as we started our work. He was safe in his generosity, for there was never any danger of my becoming affluent. Eighteen inches of coal so solidly held that every bit had to be blown from its place. Often I had to bore six holes during a shift, forcing the drill slowly into the coal by the pressure of my prone body. Behind me was a tram which held about two tons, and by the time that one was full most of my time and energy for that shift was exhausted—yet one tram alone would not earn the minimum wage. Yes, he was safe in his promise. He knew he was.

At last it has been decided to pay the skilled mining worker an added inducement to use that skill. The amount is one shilling a day higher than the labouring wage. When extra tax is deducted, his tools maintained and renewed, he will not grow very much richer because of his added responsibility and work. Yet that shilling award has brought its fresh grievance, as there was not enough money in the wages pool to give this shilling to all classes who could claim to have special skill. So several types of underground workers are without this increase and have to be persuaded that they must not slacken their efforts because someone else is recognised and they are not—and that persuading is not usually an easy thing to do.

After the agreement on holiday pay a claim was put forward for the injured and sick of the industry, who had not worked during the last year. It was felt they should have some consideration, because the industry itself had disabled them and prevented their working to become entitled to holiday pay. Their claims failed because there were over ten thousand such men in South Wales alone—over ten thousand—and the financial cost was too much.

We have the other problems of the war wage: when a man works at weekends—shall he be paid one, one and a half, or double war wage? At a certain time on Sunday the payment ceases to be double and becomes one and a half. At other periods it is one and a fifth, or again the bare shift of ordinary time. Then we have the continuous shift men who regularly work seven shifts every week. It seems that their minimum wage must be calculated for seven shifts instead of the usual six. As I understand it, for example, if four pounds ten was given as a weekly minimum for their work on the outside of a colliery, then they would have to work the seven shifts to earn that amount, and would get six-sevenths if they worked but six days. Yet supposing another workman from a different job went on the Sunday to do their work or help them, then he would get extra for that day. This extra would be double time, or time and a half, depending on what part of the Sunday he was working his extra shift. So he would get two shifts, or one and a half shifts more than the regular men, for doing the same work.

Very truly we are getting more tied up in knots with each new award. Our Federation, in despair, has now decided to print a booklet to explain all the new complications in payment. If they get the usual word-wasters and flourishers to do that little job and season it well with heretofores, henceforths, and therefores, it will be very helpful and all we will need is a second and larger book to explain the first. I am reminded of another Brains Trust session when one of the experts fumbled around with all the complicated wordings which had been stored in his brain until the Question Master explained decisively: 'He is just trying to say that he does not know.'

Our surface men, surely a total of sixty or so, have a grouse that they do not get the same allowance of extra coupons or boot vouchers which go to the underground workers. They have a deal of justice on their side, for they wear and damage quite as many, if not more, clothes. The argument from the official side is that they work outside the colliery and cannot be given extra coupons without the other industries putting in a similar claim. Probably that is correct, yet these men do suffer much of the filth and the tearing of clothes which goes in company with the handling of coal and trams.

I have been puzzling a little lately about the two largest mining villages in this valley where I have spent most of my life so far. Both are near the same size, somewhere about six thousand inhabitants. Both exist by mining, and that mining of very similar coal seams, and there is little difference in their placing in this valley. Yet life in these villages must be vastly different; I have lived for years in each village.

Yesterday I spent some hours in the one, Resolven, renewing old memories and observing well what went on. Scarcely anything had altered during the last twenty years. The shops seemed poorly served, and gave that impression of being open for the last week which is so frequent in the older mining towns. I know the people there are as decent and energetic as the folk of Glynneath, yet there seemed to be a depressed attitude about their everyday life. When late afternoon came the miners clomped home into the grey streets and homes. They brought their pit smell, their dust, and their dirt with them. Some may have washed at once. Their wives must have the water heated and ready, and when the heavy tub was carried in the kitchens were soiled with floating dust. Then the working clothes had to be dried ready for the next shift, and if men were working, from the same home, on alternate shifts this bathing and drying went on during most of the day. The absence of pithead baths had much to do with the dismalness of Resolven.

Yet there were many who lingered before their washing. Some went out to the garden working, others to the allotments, and others just lounged, dragging out a doormat and sitting on it whilst they watched what happened along the street. It appeared to me exactly like mining conditions as I first knew them a quarter of a century ago. Apart from baths, I think the greatest hindrance to a brighter village were

Trealaw, c. 1943, pencil, 30 × 27

The alleyways between terraced houses are as characteristic of valleys communities as the terraces themselves. They often lead up to the mountainsides past back yards and outhouses.

MINER'S DAY **139**

Miners' Houses, Trealaw, c. 1943, pastel, 50 × 39

When reproduced to illustrate the essay 'Coal: The National Plague Spot' in 1946, this drawing was titled 'Houses with Slag Heap'. The row of one-up, one-down cottages sits tightly under a steep-banked spoil tip; a disturbing image even two decades before the Aberfan disaster. The fence posts show signs of slumping and the thin grass covering is torn over a collapsing face of spoil.

the streets built long ago by the mining companies. There were hundreds of such houses, a solid block of the grey, unimaginative past, rushed up to provide quick shelter for the incoming workers. Those streets, abutting direct on a crooked main street, seemed to dominate and subdue all life in that area. Inside they were as comfortable as many modern houses, and I have not heard it said that the colliery companies are bad landlords, but the lack of wise planning in the past still seems to throttle all activities in that area.

Yet in Glynneath, of similar size and economic disposition, and populated by people who have intermixed and intermarried with folk from the other village, there seems a brighter outlook and a happier life. The baths here help, for many of the wives have forgotten what it was like to have a husband coming home grimy from work. Also more brick was used in the buildings, and the majority of the homes are semi-detached. As many have club houses, paying for them from their earnings, the houses are kept bright with paint and flowers, and a neglected garden is rare—almost a sure sign that it is a rented house. A little thing, just some added cleanliness and brightness in our every day, but what a total of happiness it brings to the man and his family when counted in the passing of the years which make up a lifetime!

Thinking along those lines, I saw a young woman last week who was trembling, laughing and crying at the same time. The postman had just brought her news. Not of wealth or some great success; this astounding news which had temporarily broken down her control was that she had been given a Council house. After more than ten years of huddling six in a room, and continued appeals to every possible source until she had completely lost hope, had come this slip of paper which had altered her vision of life for herself and the four children. 'Indeed, I feel as if I had been left a big fortune. I'm right now for life.' Already she was full of ideas on how to make the most of the furniture, so badly knocked about in those cramped quarters, and already her friends around there were asking should they come and see her a couple of times every week, so that they could enjoy the luxury of a bath in a real bathroom!

One of our Pit Production members was in a noisy humour and was letting everyone near know his trouble. An engine had broken down—I think the spindle was cracked. A replacement part was ready in a Midland factory, but they could not deliver for a day or two. They sent a lorry with it which broke down and was delayed for twenty-four hours. Before that our management, so this man stated, has tried to get a travelling permit and eight extra gallons of petrol so that they could go to fetch the spindle or, after a lapse of some days, that they could meet it on the way when the lorry was broken down.

'Not a blinkin' hope,' he asserted, 'and so the men had to go back home on five more days than was any need.'

Only a section of the men were affected, but they could not all be put to work in other areas. They dressed and walked to the colliery top, were told they had no place to work, and went back home each day. Six hundred and forty-six coal shifts lost, said my informant, and that would easily mean a thousand tons of coal. Through the guaranteed wage the workmen would be paid, but a very expensive piece of paper would finally have to be drawn out as a cover for the refusal of that travelling permit—and what about the coal so urgently needed?

Either the slopes up my favourite mountains are getting more steep or my wind is not what it was. I have nursed it fairly well, too, smoking very little and drinking very rarely—small amounts at long intervals. Yet I had to pause three times when climbing the mountain slope which I used to hurry across, and I was sweating profusely. I do not value my sweat to the same extent as Crush, who insists that his is worth 'a quid a spot, aye it is. Too good for any colliery company to have more than two drops a week.' The thought and sight of sweat running off my hand when I passed it through my hair when lying in bed made me forego my usual Saturday morning's sleep after a tiring night's work and sent me on a journey to Ammanford—my first visit.

Pontardawe seems to be a succession of tall smoking chimneys contrasted with a lovely church. Gwaencaegurwen is a name which always recalls the playing of a fine band, and certainly they seem to have plenty of barren open spaces and fresh mountain wind to assist their practices. Ammanford seemed a fresh and lively place, which should be a pleasant spot in which to live. Yet what amazed me was, although I had travelled no more than twenty miles and the sights I saw around me were the things I know every day, there was something so very different—after a very short while I knew what it was.

In Ammanford and its area the Welsh language was still dominant, instead of being a side issue as it is in our area. Everywhere the lovely cadences of the Welsh language pleased my hearing. Men, women, children spoke Welsh and no language but that. When I asked a question in English of a bus driver he answered dourly in Welsh. When I showed that I understood his meaning and could converse in his language he became friendly and talkative. My pronunciation was halting and my phrasing confused, but he forgave that, and I was no longer a foreigner. Four times I asked directions in English of different folk; each time I had to come back to their own language. They were quite right, these vivacious men and women in that attractive area, and I wish more of our mining areas were like that. What right have we invaders to take not only their country, but their language from them, and substitute this jumbled mixture we term our language in its place?

Tired and weary, the warmth of the bus lulled me to sleep on the way home. I was awakened in Neath, to realise that I was still in Wales, but in a part of that country which had allowed its speech and its culture to be conquered. Unusually weary and tired, I slept until Sunday morning.

I recall once travelling in a train with an old lady who was going home after her first—and as she stated most definitely, her last—visit to Wales. She did not like Wales or the Welsh folk. Why? I asked. 'They're really insolent,' she said. 'Some of them insist on talking in Welsh.' She must have been close akin to that old lady who said it was no good the Germans praying for help because God only understood English.

CHAPTER TEN

It had to be fine that day. The sun had been prayed for, watched for, hoped for, sacrificed for, yet the morning came grey and dark with a fine rain falling—the clouds were in tears because of the sun's defection. The people in that village were as gloomy as the morning, for it was their one day of the year. A few men in overalls and sheltered by mackintoshes walked dejectedly along the main streets and entered the chapel doors. Soon after, from the vestry chimneys and the outside fireplaces, the smoke billowed upwards, but was quickly defeated and sank downwards to mingle with the dampened coal dust. Women, with coats over their heads, hurried from one of the doors in the wall which represented their home to another similar door which controlled the comings and goings of a neighbour. A majestic policeman, thick-booted and caped, paraded down the main road as far as the last public house, observed that the streets and houses were all in their proper position, and returned stationwards.

The village appeared stunned, insensible to the morning. The firelighters drooped homewards, with the dragging steps of men who have wasted so much effort on a hopeless enterprise. Then towards twelve o'clock the smoke plume at the top of that high steel stack varied in its continual drift towards Pontneathvaughan. It became halting and confused, as if the breath of the fans beneath had failed. It circled, hesitant, around the slender steel stack pointing up to the sky like a giant pencil, and then the plume steadied, to drift directly towards the great bulk of Hir Fynedd mountain. Swiftly again, as it gained no welcome from that vast solidity, the smoke circled and wavered like an army waiting for orders from the centre of the stack until it decided its course, swinging right round and pointing like a moving signpost towards distant Swansea and straight out to sea.

A minute later, as if something of its wavering had penetrated into those huddled houses, a man stood in an open doorway and looked upwards. Surprised, he stepped outwards, straightened his shoulders, rolled up his shirtsleeves and, waving his hand delightedly, called to his wife to come out. As if that call had moved a lever all down that long street, a dozen doors opened as men or women came out. 'The wind have changed. See?' As they confirmed the information others came out until most doors had their watchers. 'Ay indeed. The wind have changed.'

The rain slacked to mist, and the mist moved slowly upwards. Soon the houses on the lower slopes could be seen, then the massed firs, a block of green in the lighter

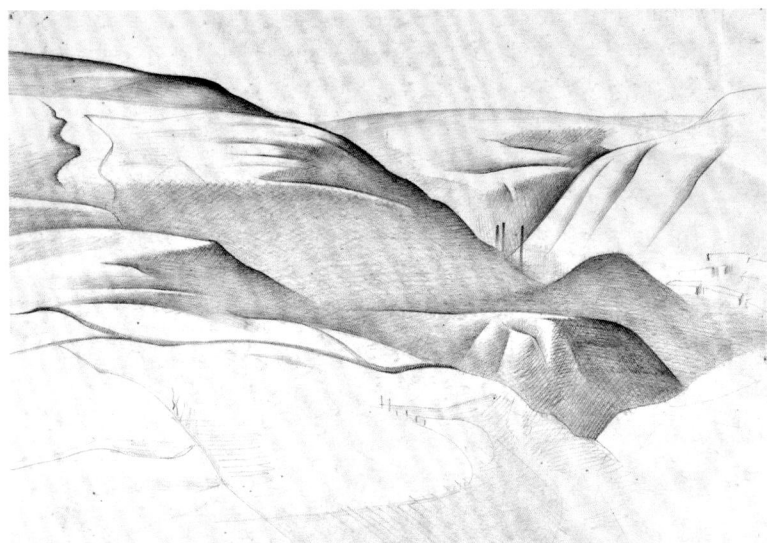

Untitled, c. 1943–5, pencil, 19 × 25

A view near Tonypandy up the side valley of the Clydach, which is identifiable by the three tall chimneys of the Cambrian Colliery. Dark waste tips push out among fields and hillsides opposite the terraces of Clydach Vale.

Tylorstown, Vivian Street, c. 1943–5, pencil, 20 × 25

The view from 27 Vivian Street, Tylorstown, the home of Jim Morton, who was one of Isabel's first contacts in the Rhondda and with whom she probably lodged. The fact that its hard-lined idiom differs from her other Rhondda drawings (with the exception of another streetscape, page 86) may suggest it was one of the first. She simplified the distant houses by removing their windows. The scene is deserted but for two children and two women washing doorsteps and pavements. A note on the back about 'lower rooms of houses back to back under road' may refer to another location.

shades of the Forestry plantings, and at last the crest of Hir Fynedd. It became moist and warm, the streets and the grey roofs steamed, whilst from inside the houses came a sound like the buzzing of angry, imprisoned bees. Men, ordered out of the way by busy housewives, lounged along the streets and collected in groups on the points where the street-ends came out bluntly against the main road. All along that street and all the other streets the wives were on their knees scrubbing the lone flagstone which fronted their own hole in the wall. The door handle was polished, the clean stacks or brown paper laid in place on the oilcloth. A smell which promised dinner came sidling from each door, now open for the passage to dry and the heat of the interior to subside; for habit is so strong that a big fire is kept in the homes of these coal-getters even when the chill of the morning had passed. The chapel

Bertie, c. 1943–5, pencil, 31 × 24

Bertie looks about twelve. He must still have been at school when Isabel drew him. The leaving age was fourteen until 1944, when it was raised to fifteen.

stokers returned to their task, walking now with the swing of men of value in the world. They found the smoke they created anew had no hesitation in following the lead set by that dominant stack plume.

Children, in various degrees of undress, started to stand at the doors, but they behaved stiffly because they knew this was no ordinary day. The mist had surrendered, following the rain back to whence it came, and then, almost unexpectedly, that for which they had been all hoping, happened. The fair curly heads, the black straight-haired heads, the golden- and the mouse-coloured heads all seemed to move together as they joined in that chorus of 'Look, mammy. There's the sun.'

After dinner, what comings and goings! Every street amove and every passage crowded. Sailor suits or pink dresses coming out of the coverings which had sheltered them since they had been bought, and what sacrifices that had often meant! Small, delighted humans in white or blue or pink, or a dozen other colours collected in groups and scrutinised one another, while the men stood uneasily about in that sombre dress which seems to be suitable only for a funeral. The wives, after a rapid 'tidy up,' were preparing themselves for the march on this fine Whit Monday.

It is the one time in the year when all our various religious sections walk together and go along the same way. At other times they claim to be going to the same destination, but each is sure the other party has missed the road. On Whit Sunday, as for generations back in this mining village, they unite in the marching and traverse every street, singing as they go. Baptists, Churchmen, Wesleyans, Pentecostals, Salvation Army, Evangelists, Missions, and the sub-divisions of each with, in such areas as these, the Welsh and English sections of most beliefs, all forget their differences for that one day. There are still some complications which must be studied each year— the problem of position and precedence in the procession often causes trouble. I have suggested that they toss for a position, but it did not seem to create any enthusiasm. There is a rotation system adopted now, I believe.

One of the sights is the banner which each chapel carries, and in the long ago a deal of interest and work was given to create and decorate an impressive banner. Often they are a carrying load for eight men, are made of silk, and have the name of the chapel and some religious wording hand-worked in coloured silks on their fronts. Waving in the breeze, and causing their bearers to stagger under the burden, they make an attractive sight—for the watchers. Many of them are of great age and are carefully guarded against damage.

In our chapel buildings can be traced the growth of the mining industry and the religious groups. Far back in the last century, when the first miners came to change this from an area which got its living off the land to one that tore its living from out of the land it was only a tiny village of whitewashed houses, each with its large garden and a few farms. The old houses still remain as a guide to what has been. This valley must have been pretty peaceful and very pretty in those past days.

When the first miners came they had their religious ideas and maintained them. They worked for them, too, often carrying stones from the river and building their own little chapels to hold perhaps a hundred worshippers. Later, with the swift increase of population, each sect became more numerous and more prosperous, so a new chapel was built and the old one retained as a vestry. That way, in several of our chapels, we have the large modern building overshadowing the ancient, which still stands solidly alongside.

There has been continual action in these vestries since the skies cleared. Young ladies, getting more attractively dressed as the day passes, arrive with burdens, pass a period inside the closed doors, then hurry home again. Children try to discover what is hidden inside, but are warned off by sombre and unsmiling deacons. The streets are a flutter of changing colours and there are crowds outside each of the many chapels. Attendances at the Sunday Schools have been large of late, probably because each child was in earnest quest of what they call 'a ticket.'

Soon after two o'clock a sound, a series of sounds, make the valley and its folk vibrant with excitement. It is the first warning thuds of the big drum as the band, all miners, starts its parade down to the meeting place. 'Colonel Bogey' lifts their feet along the grey streets and hurries the stragglers along to their meeting.

There was a time when the band led the way, but this brought in difficulties, because they marched properly and the followers did not; with the result that there were several streets between them at the end. So now the band is somewhere about the centre. There they can be kept in check by the restrained steps of the leaders.

It is not right to class this as a parade or a march—it is a saunter. Whatever chapel leads on a particular year, its most prominent men go in front. As most of these are elderly, and even the most feeble who can push one foot in front of the other will come out on that day and claim his place, the pace is very slow and uncertain. The beat of the big drum brings no message of 'left foot down firmly' to them; probably if they were to attempt it they would get into a tangle. So they shuffle along, content that the band is present for their glorification and to show the miners what they give that penny a month deduction to maintain.

Then the many cripples, or the helplessly old who manage in some way to get to the front doors, must have their greetings, too. No old friend must be passed, and by the time an old man has forsaken the lead to totter across and shake hands with some old crippled friend, and returned to his place for the walk to be resumed, quite a time has passed. This period has given an added chance for the choirs to sing, each its own song, sometimes in Welsh and most often in English. No street of even a dozen houses must be slighted by ignoring, and the route must be carefully considered with that end in view.

So they saunter and wind their way up and down the streets, singing their songs in voices which echo along the lower slopes of the valley. Not an orderly procession, but a colourful one, with the soberness of the male clothing contrasting with the colours of their womenfolk and the slowly moving groups of gently singing children whose colours and faces mingle and intermingle as, wide-eyed, they follow along.

The end comes, the band stops, and each school departs for its own chapel. The old ones go indoors, thanking God they have seen another 'march' and wondering if they will see any more. The children, tickets in hand, trot on towards those vestries. Now the doors are open, the trestle tables are white covered and laden with all the niceties which love, stinting, and the conditions will allow, and there are real flowers to contrast with those human flowers which sit so expectantly on either side.

Replete, the youngest eyes get heavy and their owners go satisfied to bed. Some of the men disdain the attraction of tea, claiming that their hobby is beer, and they are very soon in a feeling to demonstrate how marching and mining should be done. As Crush says, 'It's surprising how easy it is to put up a dozen pairs of arches in a pub. No need to blow the sides down. No bad roof to fall on you either. You should see me drawing plans in the sawdust.'

The band has accepted one of several invitations to tea, and then, with wind rather hampered, they go down to the meadow and play for sedate dancing. With dusk arriving Kiss in the Ring becomes a favourite—and kissing outside the ring also.

All who were able had watched or marched, but among those streets on that fine day, there were some who could do neither. Gwilym was one of these. I went in to see him about five o'clock, but it was too late. The blinds were down; his bed had been placed so that he could see what happened up the outside and across the houses exactly opposite. He was a great one for parades, a man born for uniform and bands, but his spirit had not survived the muggy dullness of that morning.

He had swaggered his way through the last war period and had roared his way through many of the years afterwards. Five years ago his expanse of chest had shrivelled and he could work no more. Always a fighter, he had tried to get some redress, but had failed to prove his claim. His strong voice had faded to a whisper, and there came a time when the bed had to be downstairs and he had to lie all hours propped up in that. Because of his straining his wife could not sleep, so he made her go upstairs to the other bed. 'You want rest,' he said, 'I'll be all right. Besides, you'll have to get used to sleeping by yourself.' She walked upstairs to please him, then crept quietly down into the kitchen and passed every night sitting in a chair listening, in case the gasping man would need some help she could give.

He knew what was coming, and made her promise to get the Union officials to help her and insist on a post-mortem after he had died. 'I know it's dust,' he insisted, 'and they've cheated us while I've been alive. Make them pay for me when I'm dead.'

In war and in peacetime he had to fight, but Gwilym was a battler right to his last breath. That evening I read a comment by a noble Lord during a debate on the Education Bill. He claimed, from the profound depth of his own knowledge and experience, that boys who have only an elementary schooling have less courage and cannot stand physical suffering like the boys who have a public school training. My word!

I walked alongside the canal on my way homewards. This is only a small canal, now bedded with weeds and smelling foully during hot weather. The path is no longer worn by towing horses, but is favoured by young miners who have companions with high heels, short skirts and slim legs. Being alone, I meditated on the history of that canal. A hundred years ago it was the mode of transport from this valley. When there were plans for laying the steel rails on which a train would screech its way along the canal owners and many others were terrified. A petition was drawn up which claimed that the invasion of that smoky monster would stop the birds singing, the cows giving milk, the crops from growing, and that even the trees would die. Also it stated most emphatically that the canal would easily convey all the coal ever mined in that area. Poor old barges, now just showing rotting sides through the canal mud; you would have had a busy time taking five thousand or so tons a day down that narrow waterway!

The old wharfs are idle and rotting; the old mills can scarcely be found amongst the trees and fern. Copper works, ironworks, brickworks; I see their ruins every day, and the tips from old and disused collieries. When are we going to make it compulsory that he who despoils a place must renew its beauty? Now it seems that anyone can come along, scatter horror and destruction until the chance of profit is gone, and then go away from the sight of his crime, leaving Nature to heal its grievous wounds. She does her best, and she is sure, but it takes too long.

Merthyr is not quite twenty miles from here, but I had never been there until this next week. Not feeling my usual urgent self I went there in the hope of getting some advice that would help me. There is a three-mile slope from this valley which, in past days, used to be past the ability of many motorcars but modern buses scarcely alter their song as they glide up it. Past the monotony of Hirwain we went, through some pleasant farming country around Cross Bychan,[19] and when I travel by train or bus I am as tightly pressed to the window as any child on its first ride. There was need to be cautious, for this road is an excellent imitation of a switchback, and I began to realise why, in a time of snow or frost, the Merthyr miners were worried as to whether they would be able to travel to work or to get back home—the latter being the major worry.

Nearing Merthyr the land became more barren.

---

[19] The Cross Bychan Inn, Croesbychan, was on the mountain road from Hirwaun to Merthyr.

Untitled, c. 1943–5, pencil, 24 × 31

This quick sketch records two colliery powerhouses with louvred roofs at Aberpergwm Colliery in the Neath valley: one with a square chimney on the left and on the right a more modern building of brick panels on a steel frame. A tall metal chimney stands between them. A couple of empty trams and a corrugated-iron shed are in the foreground.

Neath Valley, c. 1943–5, conté, 26 × 35

Isabel made a handful of drawings around Bert's workplace in the Neath valley. This shows the village of Cwmgwrach, nestled under the south-eastern slopes of the Neath valley, with ruins in the foreground.

Soon it was grass and tips, then tips and grass, and finally tips only. No shining sun could make Merthyr attractive; it remains as a black condemnation of our industrial policy, almost as completely bare as Landore or Llansamlet. You could read the story in the features of the men there, in that place where a King promised something should be done—before he lost his throne.

I walked into Woolworth's, scenting a cup of tea. Finding the counter at the far end, I felt in my pocket for twopence, and before I found it, or said a word, a cup of tea was handed to me. That was real service, and unusual in my experience. In other Woolworth's I have often had to wait impatiently whilst the group of young ladies who should serve the waiting crowd argue about the attractions of various face powders, or film stars, or dancing partners. Outside again I looked up at the street names; the first I read was the one I sought. Merthyr was coming out well. A man rode past me on a hunter, and he rode well, only just rising with the swing of the horse. Well breeched, well saddled and well groomed, I enjoyed that sight.

I had not seen its like for many years. After my business was ended I walked again into Woolworth's, with twopence in my hand. I was almost too late again, for the exchange was immediate. 'You again,' said the short and fair tea expert. 'Yes, again,' I agreed, 'because the first was so nice.' It was, strong, hot and even fairly sweet. A model cup of tea.

With nearly an hour to wait and plenty of buses travelling I wandered along, saw a fine library and went in. So early as that, dozens of people were exchanging books, and the scene was animated. The reading room was good also, and I glanced across the Welsh and English papers there. Lord Traprain, so it appeared, had been presenting certificates to mining workmen over seventy years of age, still in the industry. I had a certificate when I was nine for writing an essay on 'Alcohol and Its Effects.' It must have been a purely theoretical essay, surely, but I got the certificate. That was easier than waiting until I was seventy. Seventy years, and most of them started work when they were well under fourteen—perhaps twelve years of age. After at least fifty-six years in a dangerous and heavy industry they are given a certificate. About a third of it, say eighteen complete years in darkness—worse than a life sentence for murder. Anyway, I have something to look forward to if I survive another quarter of a century—I may not get any pension from my industry, but I might have another certificate to go alongside that other!

Nearing the station I saw a poster which stated that S. O. Davies and some others were to speak that night at a protest meeting against the proposed expulsion of Aneurin Bevan—a result of his speech about the 1AA strike regulation.[20] Already a small crowd was discussing it, and immediately I heard a voice I knew, two voices, three voices. Yes, they were there, William, John, Dick, Crush, Ned, and quite a dozen of the men who had worked with me the shift before in that colliery nearly twenty miles away. Now they were at home and I was the invader. There is an indescribable tie amongst men who have worked long together in danger. Here is one of their kin, a blood brother in the industrial army, a man who knows their joys and their sorrows.

I had a dozen invitations home to dinner, and I evaded them all. Anyway I did not intend to eat what their children might have needed. A drink then, and Crush especially thought this was an occasion when he could show me how to 'knock a few back.' No. I had to be home by three o'clock, so I stated, knowing well the danger of lingering both from the transport and the sobriety point of view. I had an exuberant, argumentative escort to the bus queue and a noisy send-off when the bus started. Strange, that in Merthyr that day I saw only one man in uniform. Too early in the day, probably.

I am glad that it seems Bevan is not to be expelled. I felt he was justified, so most emphatically did all my mates, and we have enough problems on our hands without being given the extra trouble that would have developed if that expulsion had been tried.

---

[20] The Regulation 1AA, April 1944, made it an offence to instigate a stoppage affecting 'essential services'. The Labour Party supported it and considered expelling Bevan for opposing it.

CHAPTER ELEVEN

George has come back after a few days' absence—like a giant, but in no way refreshed. Benjy has returned also after a short illness, and assures us that:

'There'll be something done here now that me and George have come in work agen.'

'Don't see as it makes much difference whether you're here or not,' Steve commented.

George looked at his shovel in a way which suggested there was a snake coiled around the handle. We had to work on a piece which was sloping sharply downhill. Very soon it was apparent that the little skill George had acquired in shovelling had become rusty through that short idleness. The rubbish behaved like quicksilver on his shovel.

'You want a bigger shovel,' Steve suggested satirically.

'No damned fear,' George repudiated such a need. 'This one is too big as it is.'

With his feet slipping from under him he tried vainly to edge the point of his shovel under the stones. At that angle it was a wearisome business because there was no hope of a smooth bottom.

George struggled on desperately. He jabbed with the shovel and the stones jarred it back in his hands, he pressed his foot on it and wriggled and twisted the shovel until it seemed to have forced its way under, then he strained to lift and the others slid off, leaving one small pebble on the blade. This sudden easing made George lose balance, and the small stone shot across the roadway whilst the empty shovel clanged into the almost empty tram.

'Hell,' said George definitely, watching the amazing flight of that stone, with his eyes bulging.

'Swearing now,' Steve reproved him. 'But it might help.'

George started again, struggling to work under those stones which seemed so determined to stay close to one another and on the floor. He wiped his hand across his forehead, then looked wonderingly at the wetness of that hand. Crush arrived at that moment and asked:

'Hallo. And where was you last week?'

'In a hell of a lot better place than I am now.' George replied fervently.

'Aye,' Crush agreed, 'they don't do nothing for your comfort in these here collieries.'

Yet the colliery welcomed George in its way. Back on the main roadway a long line of double timber was creaking, then cracking, then bending until the snap came. It seemed all part of a plan to increase George's uneasiness. Above us a vast army seemed moving across a shaking ceiling and the plaster—of stones—kept falling near our ears. There came a slight smear in the roof, it widened to the tracings of a crack, then it was an inch wide, two inches, six inches, then suddenly it was on the floor. Strong posts thick as a man's body and quite nine feet long were being forced down through the softer upper flooring. And as the roof came down the floor was pressed upwards. That nine-foot post began to look like seven feet, then it seemed short at five feet; but at last the post butt touched solid bottom, and then it had to break, or crush up. Some of the post tops resembled a mop. You could watch the pressure finding the weakest point and then crushing through. A knot in a post, a piece of sharp stone digging in to the timber, a wedge placed too tightly and too far away from double support—that point would snap. Sometimes that breaking point was very close to an ear which was being held tensed for a warning and a terrific 'crack—crack—crack' would deaden the ear drums. They broke as swiftly and neatly as a matchstick, but the sound, amplified by the hollow roadways, was terrific.

Our rubbish came down all right, but not when and where we wanted it. We began to think that some invisible agency was above us in the darkness, waiting until we stood underneath, so that he could drop a few stones on our behalf and in our special direction. We often waited and listened for a couple of minutes, then with everything quiet would advance to work. As we got underneath the gaping hole again, drip—drip—drip, down came the stones; so back again to safety we jumped, or at least to a place where there was some shelter above us.

The risky part of this type of work often is that the sides must be cleared before any more supporting timber can be placed. These must be put on a solid bottom and back away from the rails. Down crashes half a ton of stone; you wait for silence to assure you of a degree of safety; then hurry to clear those stones, and when almost clear the one who is watching what goes on above shouts, 'To you,' look out. By that impossible advice he means *Get back*, and you do so swiftly, just in time to evade another crashing fall. Then the whole process is repeated, and this intermittent working and jumping, cleaning and re-cleaning may go on for hours, perhaps most of the way through the shift. Very often, when one of the upper officials comes for inspection, everything goes quiet and your complaints about delay seem groundless.

'Seems all quiet to me,' one official commented. 'I don't hear anything moving.'

'It's frightened to move when you're around,' said Crush, hitching up his moleskin trousers in a way he has when he feels like dispensing trouble.

George leans dejectedly on his shovel as the stones crash down, and his face shows plainly the question 'What have I done to deserve this?'. Yet he has one bright consolation; he has told me about it on the way in. No more eating food from a stuffy food tin and drinking cold tea from a jack for him. He has made other arrangements while he has been idle. Now he has a nice clean cloth around his food and a thermos flask to keep his tea warm. 'Lovely, it will be,' he assures me; 'you ought to do the same yourself.'

'Yes,' I agree dubiously, for I have had experience of this type of ambition, 'but have you put it safe?'

'You bet,' George was sure, 'as high up as I can reach.'

Food time comes, and we prepare our seat of cold stones, place our lamps in a circle to frighten away the rats, and wait for the dust to settle before we open our food boxes. George goes off jubilant but comes back quietly. The flask in his right pocket just brushed against the rock sides, and his warm tea is finished forever. He has no jacket pocket on the left side at all, and only a few crumbs mingling with the shreds of cloth show what the rats have left of his food, although he hung it so high and so carefully. Disgusted to the limit, he sits down. Between us we collect a meal and insist on his sharing.

Benjy, hampered by a full mouth, criticised the BBC programmes. The only one he liked properly was the Happilog. Whether it is the Happidrome or the Epilogue I am not quite sure, although I cannot imagine Benjy listening to the Epilogue. Definitely he does not like 'them sympathy concerts.' After food is finished he gets up and studies the 'squeeze.'

'Jawch,' he comments, 'but it have come down. Still man's height, though.' He stands under the lowest roof and just manages to measure a couple of inches above his head without touching the top. His allusion to man's height is the habitual statement that a roadway underground must be of a man's height, which would be near about six foot, so what is anyone arguing about?—there is man's height there.

Quietly, and with deadly seriousness, George is working on some idea of his which will bring revenge on that rat which stole his food. He has got a food tin, and on top of it, placed in good sight, is a waste piece of bread and margarine. It appears innocent enough, and George believes the rats will think that well of it. Yet he has been busy working tiny chips of wood into that piece of food and he is now distantly and patiently waiting for a result. The reward for which he hopes is that the rat will creep forward and gobble that bread down. Then the chips will stick in his throat and choke him. The time passes, but we see no sign of George having his revenge. He thinks the rat has not yet become hungry again. I think he has underestimated the craft of those four-footed vermin, whose pattering feet and sharp squeals are the sounds which form a background to our working lives.

So that night passes with tense watching whilst the stones drop from above; and frantic, hurried working to clean the way and timber the openings during the quiet periods. Somehow, as is the way, we managed to get it all fairly secure, and the last hour was spent more contentedly in the knowledge that we have shelter above us and the following shift we have it also. The fireman comes along, notes that the work is well under control, says a few words which show he is at least satisfied, and goes along to another heading. Benjy comes along with his inevitable inquiry as to his whereabouts and 'What time is it?'

Then sharply into our senses cuts a shrill sound—the scream of a rat. Not the usual fighting sound, but a long call of agony. Somewhere in that crushing mass—much quieter now—one of the thousands of rats has got caught, and the roof is coming down on it. No help would be possible in there even if someone would give it. The squeal continues ripping into our hearing, then becomes fainter, and fades into silence.

'That's George's rat,' said Crush, 'he was too fat.'

'Wonder what they lives on,' pondered Steve, 'when George don't provide them with grub.'

'P'raps,' said Benjy, who was sawing himself a small block of dry wood, 'it was indigestion he had got.'

George was silent, less immune to his surroundings. After a while he spoke his thoughts. He said:

'It might have been a man. It might have been me.'

Somehow the shift seemed harder and longer that morning. My chest seemed to burn as if I had swallowed hot dust. Outside the sky was blue, with the sun coming up bright and warm in promise of a lovely day. We stood and let the loveliness of it invade our minds and bodies. Later, clean and refreshed by the shower bath and with ordinary clothes on, we met again and saw each other as ordinary human beings. The first one in carried six cups of tea on to the small white table and we munched biscuits, a rare treat, whilst we chatted.

The clock showed its warning of passing time, and George, with broad shining face, went out; then Crush with his wide-shouldered swagger; Steve with twinkling eyes and laughing mouth, and last a hurrying Benjy with the crumbs still on his lips and his short legs moving swiftly. As they went, each repeated the farewell of:

'So long, butty. See you next shift.'

Next shift. Tomorrow and all the tomorrows.

■ ■ ■

For a long time, I have been sitting and thinking. No one is near me and there are hardly any disturbing sounds. A couple of seagulls have just passed over my head, screaming their derision at the sharp breeze which sings through the short, dryish grass on the mountainside, and lifts the moss into the holes of the rough stone walls. I am in a sheltered hollow which might have been a quarry in the long ago before time turned the greyness back into the soothing green. The wind passes above me without ruffling my hair, the day is warm, and from a full mile away I can feel the shudder of the air as the great compressors draw it inwards. That sound, softened by distance, is all that disturbs the peace.

I know this spot well; I have been here before in the periods when I want solitude. For that purpose it is ideal. Slight oak trees have grown up from the hollow, and some crooked ash. The sun passes amongst their leaves, outlining the whiter veins on their surfaces and making the hollow a place of sunlight and shadow. The brook moves slowly below, and when the sun catches it the water flashes its delight—short-lived pleasure, because the ground drops from underneath it and the water slides, it is not a direct fall, for about sixty feet. It is now a silent slither, because the brook is almost dry and the stone sides of the waterway shine with the polish of continual friction.

Near me, past the huge and almost covered stones, is the ruin of a farmhouse and its buildings. The garden path is still traceable as the stone edging peeps out frequently. A large white rose on its standard holds itself upright and aloof from the straggle of wild roses which try to hide the fallen stones of that abandoned homestead. The honeysuckle has joined with the wild rose in its effort at concealment, and their fragrances scent that lonely place. The white bloom of the hawthorn has surrendered to the season, and now lies like perfumed confetti around the places where the honeysuckle and the wild rose mingle in their lasting lover-like embraces. The white rose, reminder of that care and cultivation which was abandoned, nods slightly like a disgusted virgin. I am walking over the hopes and defeats of other people. Very surely they did not wish to leave this peaceful spot; but less than a hundred yards away, in a deeper hollow, is the thing which drove them away and surrendered their land back to the uncultivated mountain.

In that hollow is a coalmine—now also abandoned. The yellowy water still flows out under the stone arching. I always wonder what vast reservoir of water must be in the earth. Steel ropes hang limp and lifeless, electric cables are growing mouldy, a pile of horse manure, another of sawdust, another of horseshoes, and a vast nettle-entwined mixture of broken steel arches and damaged tram wheels show how busy this scene once was. On one brick building the First Aid notice still shows plainly, and another explains its use as a Lamp Station. Extracts from the Coal Mines Act front another large building, and row after row of empty coal trams stand in their dozens on the rusting rails.

Untitled, c. 1943–5, pencil, 28 × 36

This view across a wooded gorge appears to be of the Empire Colliery's spoil tip on the hillside above Cwmgwrach, which was served by an incline. Isabel would have had to walk steeply up from the mine entrance to see it.

Trucks, c. 1943–5, pencil, 24 × 32

A graveyard of derailed coal trams.

That way they have stood for more than three years. The mine drove the farmers away, then after a brief busyness the miners walked out from their last shift. Idle works and abandoned farms—that seems the history of our valley. Yet during the last three years the country has been calling for coal and for what the farm could produce. Wealth on the land and from under the land neglected when our need is urgent; and space, with fresh, sweet air flowing around us on every side, while down below in that valley, from which the call of hawkers comes up to us gentled by the distance, are thousands of decent human folk inhaling dust each time they eat or breathe.

'I suppose you are working in dust,' asked our doctor, when I went to see him yesterday about that tightening, burning chest.

'No more than I can help, doctor,' I assured him.

'No. I expect not. But how can you avoid it in these places, where it follows you into your homes? I wonder the kids in school don't get it.'

He wrote and sealed the letter, then handed it to me, saying:

'You know where to take this.' Then he leant on his desk and looked out on the mining street. From the waiting room came the continual rise and fall of sound as men and women waited—and some had to wait for long periods.

'It's about getting me down,' he spoke slowly. 'Men that you have known for years coming to you and you can't do much for them. And you know how little they will have to live on in the end. Well, good luck. I know you will be able to handle your end. Wish they all could.

God. It's damnable.' He pointed to a pile of open letters. 'That's this morning's problems—human problems.'

In our areas we seem outside the range of guardian angels. We replace them, because our need is severe, by union officials and doctors. I know well that there are misfits in each section; that some union officials guard themselves first, second, and last, but they are the minority and the men usually find them out. I know also that there are doctors who betray the name, who are brutal and overbearing in manner, but they are also in a minority, and for most of the mining doctors no praise could be overdone.

Often alone, in their profession, at a village with some thousands of people; with fresh injuries coming to them each day they have, night or day, no chance of real respite. Usually a crowd is waiting the opening of their surgery door, and as one is treated two more arrive. The crowd overflows into the street, and surgery time frequently lasts four hours. Perhaps an urgent message is rushed in and the doctor hurries away, while the crowd settles down to wait for his return. He must always be at some bedside, yet always be at home. Passing his hours amongst sickness, he must never be ill himself. There at a birth, and at a death. In the life that intervenes he is confessor, clerk, and general adviser to his people. When depression stops the pay coming to the workers his money stops coming also, yet his work must go on. No Harley Street or palatial home for him; his abode is an advanced dressing station in the everlasting war against industrial disease and accident. We need much more help and encouragement for these doctors, and better facilities. I think, too, that a woman doctor should be in every sizeable village.

A couple of hours later I was again waiting my turn in another room. Not so crowded, but still being constantly visited, and another man almost always coming in when the front door closed behind one going out. All men, we sat around grimly silent, and each knew exactly why the others were there.

'Next up when the bell rings,' the notice warned us. My ring came and I went upstairs, meeting the last man coming down. 'Name, please.' I gave all particulars, then stripped to the waist. I felt slightly amused because one of the doctors was explaining a cure for stammering to a man who was replacing his clothes. I felt that in such a case it seemed a waste of good intentions. I tried to see if my body had shrunken as I held my breath while standing before the X-ray screen. Is exceptionally keen hearing with a knowledge of physiology and anatomy always a desirable attribute? Even at that distance I could overhear the discussion. 'It looks to be here—and over here. We had better take another for surety.'

An hour later I was sitting in the Victoria Gardens at Neath. This is a lovely little park just away from the busy streets. The flowers delight the eye and above it all the clock in the tower of St David's Church warned us of the passing of each bright hour, and the sun was genial that day. Out in the warmth the people moved about their lives. The shops were busy and loaded lorries whined past. This was normal life to them, the only way they understood. What did they know of roof falls, or stiff working coal, or men who gasped as they breathed?

Washing, 1944, lithograph, 8 x 9

An original illustration for *Miners Day* showing laundry on a line slung between a terrace and a timber pole. The houses have porches made of rough timber and corrugated iron.

What did they care while the sun shone and the world was bright? Yet nearly three-quarters of a million men—my mates—were shut away from that sun. Their normal life was about them then, in a crushing darkness with sweat running down their back to make their singlets like a wet cloth. Men straining to rip out the compressed sunlight which had been stored in the heart of creation uncounted centuries before; and which to a greater or a lesser degree affects all our lives and our national prosperity.

I felt weary suddenly, out of place. I wanted to go home, and a bus was waiting. I watched the outposts of the sea just licking at the unused wharf; and away in the distance could just detect the outline of Neath Abbey, a ruined monument to the greatness that survives after eight hundred years. As we travelled down the Neath Valley the mountains showed very close and clear and the forestry plantations on the lower slopes waved gently like a huge, and dark, green fan. The grass was lengthening, tinted with buttercups, and the swedes were coming out in long, thin rows of tiny green dots.

The grey blotch of Resolven affronted the beauty in that width of valley. Farther along the rhododendrons were massed in flower, like scarlet soldiers alongside the lake at Rheola. Every familiar sight seemed to come out clearer to my mind that day; somehow I seemed to be watching for something which I could not define. With my mind wandering, I suppose, I got off the bus at the Cwmgrach halt instead of riding to the next stopping place.

A hundred yards along that road the taste of dust reminded me that the colliery screens were at work. My eyes were gritted and that acrid, bitter feel came to my palate. Can you recall some evening when a sharp shower fell on a roadway that was deeply laden with dust? It damps the dust but does not soak into its depth. As you walk the disturbed and weighted dust seems just able to assail your nose and mouth without going any higher. That was the taste I met on the road. It sort of holds your breathing and dries up your tongue.

At the first turn I branched away and started climbing slowly, for I have already learned the need to slacken a racing speed. For a while I did not wish to go homewards. I wanted to get easily across those slopes and into that hollow. Soon the dust and noise was below me. The smell of trodden turf seeped into my sense and drove the dust thoughts away.

Then, with the highest point in Glamorgan clear above me, and a bedraggled mountain sheep with a young lamb which bounced about like a four-legged ball as my only companions on the crest, I found my hollow again; and here I have been sitting—thinking.

Untitled, c. 1943–5, pencil and ink, 28 × 21

Run-down farms stood on the hillsides above the colliery townships of the Rhondda. The slate-roofed house has old corrugated-iron sheds against it. Isabel made colour notes: 'Door and window here deep ochre. Roof light, lighter than hill behind. House on right very black indigo, shining in rain.'

Is it the ignorance or the indifference of man which permits this senseless and endless waste of men, of land and of minerals? In such a smiling land why should there be so much misery and so many grieving minds? With such a generous Creator—for there are coal deposits, iron stone and clay by the thousands of tons beneath me and miles of open and unused land all around me—why should folk go short of food and comfort?

A wise control would use all this for our good. Instead, as I know, the coal has been left in the seams, the farms have gone derelict and hundreds of men have lounged, as I have, in bitter memory. Such a system does not deserve

Untitled, c. 1943–5, watercolour, 36 × 31

These houses were probably in the isolated rows around the Glenrhondda Colliery and its spoil tips above Blaencwm. They may be the splayed mid-nineteenth-century rows known as 'Stone Huts'.

to endure—it cannot last. Yet high above me, the sky is blue and serene as if mocking us with its vast peacefulness. A wisp of white cloud moves distantly across as a single life passes across the span of our time. The brook glides down the slide, the leaves defeat the efforts of the sun to look through, and the shadows get wider and longer. The grass feels damp; it is time to go home.

Farther down the valley sounds come again to meet me. Children playing by the warm riverside—from a long way off the sharp pips which precede the nine o'clock news and the tang of wood burning on a house fire … and that fire in your own home.

Then, when we have asked the same hesitating questions which have been asked so many times in this area and given some evasive answers—'Will have the report in a few days. Feel all right so I expect it will be OK.'—we realise that long evening is surrendering to the dusk.

Outside again I walk across to look down into the valley. Hir Fynedd seems huge and very close. The sky is clear with a few stars pointing their light. The distant compressors pant and shudder as they gulp in energy for more work. Is it again that time? Yes. The staccato warning of the whistle came as the brakes started to ease the speed before entering that station. Benjy and Steve would be on that train with all the others. I saw several white blobs of faces looking up towards me and a couple of hands waved. Not tonight, boys. A wet nose rubs against my right hand; the dog has sensed my feelings and hopes to comfort me. I stand quietly until the other train rushes past. George, with Hermit and Crush, will be there on those hard seats. Getting started for another shift in which I will not help. So Goodnight, workmates … and perhaps, Goodbye.

GLOSSARY

Some terms varied between coalfields; the definitions below relate to South Wales usage or the sense B. L. Coombes intended.

Airway – an underground tunnel that acts as part of the ventilation system for a mine.

Anthracite – the hard, pure coal that is found in the western part of the South Wales coalfield, used especially in malting and later in enclosed stoves. Scarring of the lungs by sharp anthracite dust exacerbated pneumoconiosis.

Anthracosis *see* pneumoconiosis.

Bevin Boys – conscripted young men who were selected by ballot at the ages of 18 to 25 to work in mining instead of the armed services in order to raise coal production for the war effort. The programme began in 1943 and was named after the Minister of Labour, Ernest Bevin.

Blackleg – a miner who continues working during a strike.

Butty – a working partner or friend. The use of the term often implied the relationship of one collier to one butty, on whom he might be dependent for his safety.

Cog – a square or oblong pier to support the roof, made of square-cut cogsticks and timber wedges packed with stone.

Collier – a miner who works cutting coal at the face.

Compressor – an engine that produces compressed air supplied to different locations underground through pipes to power pneumatic tools and machinery.

Curling box – a flat, three-sided scoop made of sheet metal with handles, used at to coal face to gather up loose coal and slide it along the floor.

Conveyors – powered conveyor belts to carry coal from the face to trams or in some mines directly to the surface.

Deep – a sloping tunnel or roadway.

District – an identified area of the mine underground.

Drift – a tunnel driven into the mine from the surface.

Fireman – a safety officer responsible for inspecting the mine for gas, dangerous roofs and other hazards and for shotfiring and use of explosives underground.

Haulier – a miner responsible for the movement of coal trams.

Heading – a tunnel or roadway underground.

Jacks – metal containers for tea or water carried underground

Journey – a train of trams linked together to be moved along a railway.

Level – underground workings accessed by a tunnel rather than a shaft.

Lid – a piece of shaped timber inserted between the top end of a pit-prop and the rock.

Locking bar, Tool bar – a bar for hanging a miner's personal tools and securing them with a padlock.

Mandrel – a collier's pick, varying in size and design depending on use.

Mining Association – the organisation representing employers in the coal industry.

Overman – a mine official reporting to the under-manager.

Parting – a siding for drams on an underground railway.

Pneumatic pick – a small pick operated by compressed air for loosening coal.

Pneumatic borer – a drill operated by compressed air.

Pneumoconiosis, dust – in coalmining, a disease of the lungs caused by inhaling carbon dust, causing pain and acute shortness of breath, and sometimes leading to necrosis of the lung tissue. The severity of the disease is related to both the extent and the duration of exposure. The number of cases increased after the introduction of mechanized coal cutting.

Refuge, Refuge hole – a small space created off the roadway underground where miners wait while trams go past.

Rider – a man who accompanies a journey of drams, riding in one of them.

Rippings – sections of roof that have to be removed before a heading can advance.

Roadway – the access tunnel into underground workings.

Screens – the plant where coal is sorted on mechanical screens to remove dust and grade lumps into different sizes.

Shot hole – a drilled hole with explosives inserted to break up coal or rock.

Silicosis *see* pneumoconiosis

Sledge – a sledgehammer, used to break up stone or coal directly or with iron wedges.

Slip – a joint in a coal seam, or the volume of coal between two joints.

Slope, Slant – an underground roadway into the mine sloping downwards, usually with a haulage engine at the top to move trams on the incline.

Snap – food taken undergound to eat during a shift, usually in a snap tin.

Sprag – a short length of wood roughly pointed at each end to push through the spokes of a tram wheel and lock it.

Stall – the working area in the pillar and stall system of removing coal from the seams whereby solid pillars of coal were left to support the roof while the coal was removed around them. By contrast, the Longwall system involved the removal of all the coal while the roof was supported by timber and packing, in some cases being allowed to collapse behind the working area.

Sylvester – a device of ratchet and chains patented by Walter Sylvester in 1895 to allow pit props to be pulled out from a distance, avoiding the risk for the miner of knocking them out with a sledgehammer.

Tool bar *see* Locking bar

Top – the rock surface on top of a coal seam, the roof of underground workings.

Tram, Dram, Tub – a mine wagon used for conveying coal on rails, underground and on the surface.

Tub – *see* Tram.

Wedge – a steel wedge hammered into the coal or rock to split it.

ACKNOWLEDGEMENTS

The editor wishes to thank Robin Alexander, Alun Burge, Charles Burton, Ben Curtis, Richard Davies, Richard Edwards, Olwen Fowler, the late Hywel Francis, John Geraint, Andrew Hawke, Carolyn Hitt, Daryl Leeworthy, Simon Miles, Lynne Moore, Adrian Owen, Joe Pearson, Dai Smith, Chris Williams, Ellie Dawkins and Karen McKinnon of the Glynn Vivian Art Gallery and staff of the National Library of Wales and the Richard Burton Archives at Swansea University.

The photographs of the Coombes family by Bert Hardy, pages 11, 13 and 168 are supplied by Getty Images ©.

All other images are © the Estate of Isabel Alexander. Bridgeman Images manages copyright on the Estate's behalf for those on pages 4, 20, 21, 22, 23, 26, 48, 51, 88, 101, 109 and 111.

The portraits on pages 14, 28 (left), 79, 107, 117 and 118 (right) are in the collection of the Glynn Vivian Art Gallery, City and County of Swansea.

We are grateful to Robin Alexander for permission to reproduce the works of Isabel Alexander and for his help with sourcing, cataloguing and biographical information.

Original image sizes are in centimetres.

INDEX

Aberdare **98**, 127
Aberpergwm 9, 14, 31, 114, **149**, **168**
accidents 50, 82, 89-90, 93, 102, 106, 116, 127, 130-1, 156
air, airways. *See* ventilation
Aldridge, John 41
Alexander, Donald (husband of IA) 16, 17, 25
Alexander, Isabel **15**, **17**
   family 15, 16
   training 15
   employment 15-16, 41-2
   marriage 16
   travel 15, 42
   visit to Coombes 10, 18, 26
   style and influences 15-16, 20-2, 25, 41-2
   publications 24, 41
   exhibitions 41-2
   death 42
Alexander, Robin (son of IA) 15, 16
Americans 68, 84
Ammanford 142
An Architect of Nature: Being the Autobiography of Luther Burbank 77
anthracite 9, 52, 99, 128, 131, 160
anthracosis. *See* pneumoconiosis
armed forces 58, 63, 68, 96, 120, 126, 150
Artificial Limb Fund 91
banners 145-6
baths, bathing 10, 33-4, 49, 75, 81-2, 89, 90, 91, 93, 98, 99, 103, 118, 125, 139-41, 154
   levy on wages 32, 122, 125
   at home 112, 139, 141
Bawden, Edward 41
BBC 81, 153
Belgians 84
Benjy (co-worker) 32, 47, 51-7, 60, 65, 74, 75-6, 80-3, 86, 91-3, 114, 121, 124, 134, 151-4, 159
Benny (friend) 89
Bertie (portrait) **145**

Bevan, Aneurin 150
Bevin Boys 38, 41, 99, 119-23, 160
Birmingham 15, 87
blacklegs 32, 63, 106, 160
Blaencwm **5**, 17, **17**, **21**, **23**, 25, **34**, 35, 38-9, 41, 48, **51**, **65**, **70**, **101**, **107**, **108**, **109**, **112**, **127**, **159**
Blaengwrach 9
Blaenllechau **35**
blasting 105, 137
Blind Institute 91, 130
Board of Trade 61
Boys' Club Movement 122
Brains Trust 138
brass bands 11, 37, 115, 142, 146-7
British Restaurants 90
Bromley County School for Girls 15
Brooks, David B. 42-3
Burton, Charles 25
bus transport 12, 47, 58, 63, 78, 81, 85, 91, 97, 125, 130, 142, 148, 150, 157-8
canteen 10, 33, 50, 63, 77, 81, 83, 89, 91, 122-3, 125
cats underground 56
chapels 66-8, 83, 106, 133, 143-7
children **11**, **13**, **40**, 67, 68, 89, 112, 122, 126-7, 145, 146-7, 148
   portraits **20**, **37**, **38**, **39**, **40**, 40-1, **95**, **118**, **127**, **145**
   play 6, 40, 93-4, 131, **144**, 159
   food, nutrition 43, 68, **127**, 150
   poverty 9, 36-41, 93-6
   work 74, 113, 130-1
choirs 28, 37, 68, 115, 147
cinema, films 66, 68, 115, 12
class 7, 13, 69, 78, 130
Clydach Vale **144**
coal 9
   house coal privileges 32, 61, 64, 72-4, 86, 109, 128, 130, 134
   scavenging 68
   reserves 90-1
Coal (NCB magazine) 21, 42
coal cutting

   by hand 9, 27, 54, 104, 135-6
   by machine 9, 27, 131, 135-6
   in varying conditions 27, 84, 102, 136
coal industry
   employment 7, 9, 10, 119
   output, production 74, 85, 90-1, 93, 115-119, 136, 141-2
Coal Mines Act 155
coal preparation. *See* screens
coal seams 52
   high 104-5
   small 52, 104-5, 128, 134-5, 137
   geology of 27, 54, 85
Coal: The National Plague Spot 21, 22, 29, 30, 43, 48, 67, 109, 110, 111, 112, 140
collieries
   Aberpergwm 9, 14, **149**, **168**
   Blaencwm Level 17, **51**, **70**, **101**
   Britannic Merthyr 31, **36**
   Cambrian **144**
   Empire 9-10, **10**, 32, **155**
   Gelli-faelog **94**
   Glamorgan 31
   Glenrhondda, Hendrewen **17**, **21**, 38, 51, **65**, 159
   Glynogwr 31
   Nantgwyn **19**
   Naval **6**, **16**, **18**, **19**, **29**, 45, 94, 110, **124**
colliers 9, 10, 84, 90, 93, 136
colliery companies 72-3, 86, 106
   officials 62-3, 69, 72-3, 83, 91, 98-9, 119-21, 130, 136, 141, 152
Communist Party 12, 16-17
compensation 27, 32, 58, 60, 64, 78-80, 89-91, 95-6, 128, 131, 148
compressors, compressed air 121, 123, 154, 159, 160
Conciliation Board 72
Connolly, Cyril 12
Controller 93
convalescent homes 123
conveyors 135-6, 160
Coombes, Bert 7-8, **11**, **14**, **168**
   early life 8-9, 11

marriage 9
children 9, 13, 42, 115-16
Welsh language 142
habits 66, 123, 142
union work 14, 50, 64, 80, 98-9, 128
mining career 9, 13, 14, 98, 130, 135-7
unemployment 11, 13, 158
farming 13, 14, 42
injuries 42, 54-5
health 142, 156-7
writing 7-9, 10-14, 25, 42, 150
published works 11-13, 24
death 42
Coombes, Mary (Mary Rogers, wife of BLC) 9, 13, **13**, 14, 42, 66
Coombes, Peter (son of BLC) 42, 63, 66, 118, 126, **168**
crushes 27, 52, 69-70, 92, 99, 152-3
curling boxes 128, 160
Curtis, Ben 43
Cwmgwrach, Cwmgrach 9-10, 13, 28, 33, 87, 114, **149**, **155**, 158, **168**
Cwmparc 18, 29
Dan (friend) **26**, 27-8, 59
Dave (co-worker) 47, 53
Davies, S. O. 150
Degenerate Art, *Entarte Kunst* 15
Depression era 25, 96, 127, 156
Dinas 16, **18**, **29**, **67**
disability 59-60, 64, 72, 78-80, 89, 103, 111, 134, 138, 146. *See also* pneumoconiosis
dockets. *See* payslips
doctors 28, 32, 59-60, 80, 89, 91, 95-6, 128, 130, 156-7
documentary films 16, 24, 40, 122
drama 28, 37, 115, 122
dust 27-31, 52, 57, 58-60, 66-8, 87-9, 102-3, 112, 114, 128, 158
damping, spraying 84-5, 128, 135
dynamite. *See* explosives
Eardley, Joan 40
Edmonds, Michael 41
education
community 89, 122-3

schools 41, 94, 118, 148
colleges 123-4
Education Bill 148
Edwards, Ebenezer (Ebby) 75
electricity, electric power 14, **94**, 100, 102, 106, 123, 124, 132, 155, 149
enginemen, engine drivers 10, 27, 50, 53, 64, 85, 100, 115, 123-5, 130
engines, engine houses 10, **21**, 25, 27, 47, 49, 50, 53, 64, 85, 100, 115, 123-5, 130, 136, 141
Evans, Tom (collier) 32-3, **33**, 87-91, **88**
Evans, Vincent 25
explosives, dynamite 84, 137
eyesight 33, 88, 116, 123
factory work 37, 56, 78, 96
farms, farming 9, 42, 94, 97, 105, 118, 146, 154-6, 158
farm animals 29, 34, 94, 97
Ferndale 35
firemen 51, 56, 60, 69-72, 76, 80-1, 153, 160
First World War 16, 82, 97, 108, 147
fitters 124
food 68, 133
underground 27, 56, 71, 83, 93, 125, 137, 152-3
canteen 81-2, 83, 90, 125, 154
Foot, Robert 120-1
Forster, E. M. 7
Fred (Trealaw, portrait) **118**
Freedman, Barnett 24
French miners 84
funerals 27, 58-60, 116
Future Books 1: Overture, illustrations by IA in 21, 22, 24, 29, 30, 43, 48, 67, 109, 110, 111, 112, 140
gardening **21**, 35, 58, **67**, 131-3, **132**, 139, 141
gas 54, 69, 105, 137,
George (co-worker) 26-7, 47, 51-7, 61-3, 69-71, 76, 81, 100, 100-4, 134, 151-3, 159
Gilfach Goch 18, 21, **22**, **30**, **31**, 31, 35, **36**, **48**,
Glyn (portrait) **95**
Glynneath, Glyn-neath 10, **92**, 94, 122, 139, 141
Government 24, 75, 127n, 130-1
Graham (portait) 20, **20**

Greene Award 73, 86
Grenfell, D. R. 12
Gwaencaegurwen 142
Gwilym (co-worker) 28, 147-8
Hardy, Bert **11**, **13**, 92, **168**
health insurance 55, 58, 116
helmets 33, **33**, 52, **118**, 119
Herman, Josef 25
Hermit (co-worker) 47, 75-7, 80, 106, 159
Hir Fynydd 143-4, 159
Hirwaun 148
holidays 71
holiday homes 123
holiday pay 80-1, 136
Horner, Arthur 12, 32, 75
horses 52, 56, 57, 84, 85, 148, 149
hospitals 91, 120, 130, 157
house coal privileges. *See* coal
housing **5**, **17**, **18**, 34-6, **34**, **36**, **86**, 96-7, **107**, **108**, 110, **112**, 106-14, **139**, **140**, **144**, **159**
company 36, 106, 141
club 141
council 36, 141
rents 109
sanitation 36, 141
temporary 36, 108-10
shared 36-7, 106-8, 112-13
How Green Was My Valley 12
Impington College 123
inclines 10, **19**, **21**, 25, 49, **65**, 98, **113**, 115
income tax 32, 64, 80, 86, 130, 137
injuries 32, 37, 64, 71, 82-3, 89-90, 127, 137, 156
insects 81, 97n, 103
International Surrealist Exhibition 15
Italians 84
Jerry (co-worker) 27, 58-60
funeral 58-9
Joan (portrait) **39**, 41
Job (miner) 82-3
John O'London's Weekly 66
Jones, Bill 8, 42, 43
Jones, Lewis 12
Jones, Will (miner) 51, 116

MINER'S DAY **163**

Labour Exchange 60, 78, 120
Labour Party 24, 150n
labourers 51, 66, 105-6, 120
Lamey, Jim (miner) 64, 103
lamps 49, 51, 56, 62, 69-70, 73, 75, 76, 98, **98**, 102, 121, 125, 127, 135, 153, 155,
Lane, Allen 24
Left Book Club 12
Left Review 11
Lehmann, John 12, 24, 43
Leicestershire coalfield 136
libraries 66, 122, 150
literacy, reading 47, 50, 66, 75-7, 98, 122, 150
Llewellyn, Richard 12
Lloyd George, Gwilym 127
London, Jack 77
López Ortega, Ramón 43
Madley 9
malnutrition 16, 37-40, 43, **127**
Manassas: A Novel of the War 77
Manton, Joseph (father of IA) 15
Margaret (portrait) **38**, 41
McKenzie, Janet 41
Means Test 60
Medical Boards 60, 72, 91, 95-6
Merthyr Tydfil 148-50
Meston, Lord 12
miners
    comradeship 26, 49, 150
    craft 26, 27, 77, 102, 119-20, 136-7, 151
    pit sense 26, 52, 101, 116, 119
    clothing 97, 118, 125, 138
    health 27-8, 58-9, 89-90, 131, 138
    deaths 27-8, 56, 58, 72, 80, 93, 102, 105-6, 116
    fines 73
    light duties 60, 78, 103
    boys 93, 119, 125, 130, 136
    young 49-50, 115-24, **117**, **118**,
    older 50, 77, 90, 116, 150
    retirement 77-8, 112-14 123
    sickness at work 59, 83, 90
    travel to work 47, 50, 75
    See also accidents, injuries, colliers, enginemen, fitters, firemen, labourers, overmen, repairers, riders, surface workers, skills, tools, injuries, pay
Miners Day 8, 9
    publication **8**, 24
    reception 24, 42-3
    original illustrations 8, **14**, 18, **26**, 33, **37**, 57, 61, **88**, **98**, **124**, **129**, **157**
Miners Federation of Great Britain. *See* union
Miners' Rest Home 14
Miners' Welfare Commission / Committee 9, 33, 121-3
mines
    heat 84
    water 84, 103, 130
    noise 135
    old workings 103, 126, 130-1
    closure 155
Mines Act 74, 155
Mining Association 120, 160
mining communities 9-10, 33-4, 37, 138-42, 143-7
Morgan (miner) 78-80
Morris, Cedric 25, 41
Morton, Jim 17, **144**
music 122. *See also* brass bands, choirs
National Coal Board, NCB 16, 21, 42
National Council of Labour Colleges 11
National Union of Mineworkers. *See* union
nationalisation 43
Neath 12, 28, 78, 97, 130n, 142, 157
    hospital 28, 157
    livestock market 97
    Victoria Gardens 157
Neath Abbey 157
Neath Canal 10, 68, 148
Neath Guardian 12, 42
Neath, Vale of 9, **10**, 12, **14**, 28, 87, **149**, 157-8, **168**
New Objectivity, *Neue Sachlichkeit* 15, 22
New Writing 11, 12, 24
News Chronicle 93
nystagmus 33, **88**
Onllwyn 42

output, production 74, 85, 90, 93, 115, 119, 136, 141-2
overmen **48**, 49-50, 56, 60, 83, 98-9, 106
parades 28, 145-7
Parish Relief 58, 60, 77
parks 114, 157
pay 32, 48, 53, 60-5, 71-4, 76, 77, 80, 83, 85-6, 91-3, 129-30, 137-8
    piece rates 32, 104-5, 128-9, 137
    deductions 122-3, 129-30, 146
    minimum wage 32, 60-1, 99, 137
    payslips, dockets 32, 60, 89, 129-30, **129**
Penguin Books 8, 24, 41, 77
pensions 77, 83, 114, 150
Penygraig **6**, **16**, **18**, **19**, 29, 35, **46**, **94**, **110**, 132
Picasso, Pablo 15
Picture Post 12, 13, 168
pithead baths. *See* baths
pneumoconiosis 26, 27-8, 43, 58-60, 89, 93, 102-3, 128, 131, 156-7
    sufferers 36-7, 58-9, 79-80, 95-6, 124, 159
    death from 58-9, 147-8
Poles 84
police 16, 89, 119, 143
pollution 29, 93-4, 148, 156. *See also* dust
Pontardawe 142
Pontneathvaughan, Pontneddfechan 94, 143
Porter Award 32, 53, 60-1, 64-5, 85-6, 99, 129-30
Porter, Lord 32, 53, 64-5
Porthcawl 12, 14
Priestley, J. B. 12
Pritchett, V. S. 12
Production Committee 74, 76, 93, 141
pubs 11, 66, 68, 90, 106, 123, 133, 147
pumps, pumping 85, 103
radio, wireless 50, 60-1, 66, 75, 93, 126
Ragged-Trousered Philanthropists, The 77
rats 53, 97n, 103, 153-4
reading. *See* books
Rees, Johnny (engineman) 124
refuge holes 49, 101, **101**, 161
religion 50, 59, 68-9, 93, 133, 145-6

repairers, repair work 9, 10, 55-6, 69-72, 98-100, 103-6
Resolven 9, 13, 33-4, 122, 139-141, 158
Rheola 158
Rhondda valleys 11, 16-18, 34, **86**, **132**. *See also* named places
Rhys Griffiths, Archie 25
Richard Burton Archive 42
Richardson, Joshua 130
riders 64, 72, 73, 81, 83, 85, 100, 127, 161
roof falls 27, 50-1, 55-7, 62, 64, 69-72, 84, 101-2, 106, 126, 135, 151-3. *See also* crushes
roof supports
    timber 27, 55, 61, 69, **70**,
    steel 27, 50, 61-2, **61**, 101, 103-4
roof testing 102, 105
Rotha, Paul 16, 24
Royston (Blaencwm, portrait) **38**, 41
Saffron Walden Teacher Training College 41
Sankey Commission 131
screens, washeries 10, 21, **46**, 63, 65, 87, 91, **92**, 158, 161
Second World War 8, 12, 16, 20, 24, 25, 41, 47, 52, 71, 120, 138
    invasion of Europe 41, 126
shift work 9, 57, 98, 115, 130, 138
signalling 85, 100
silicosis. *See* pneumoconiosis
Simons, Dick (miner) 83-4, 150
Sinclair, Upton 77
singing 53, 68, 145, 147
Slade School of Art 15, 16
Slaughter, Alf (miner) 116
Smith, Dai 31
smoking, tobacco 49, 53, 87, 98, 103, 142
Sochachewsky, Maurice 25
socialism 133
South Wales Miners' Federation. *See* union
Southey, Robert 114
sport 60, 122, 127
Stephens, Thomas 94
Steve (co-worker) 26-7, 32, 47, 51, 53-7, 77, 81-3, 87, 102, 121, 134, 151, 154, 159
Story of Plant Life, The 24, 41

Strachan, W. J. 96
Strand Films 16
strikes, stoppages 32, 62-4, 72
    1926 9, 32, 106, 134
    damages 32, 134
Studio, The 96
Sunday closing 68-9
surface workers 129, 138
Swansea 20, 143
telephone 76
    underground 127
Terry (Blaencwm, portrait) **39**, 41
These Poor Hands 12, 13, 24, 25, 32
Thinker's Library 77
Thomas, Gwyn 11
tips 6, 9, **16**, **18**, **19**, **21**, **22**, **23**, 25, **29**, **30**, 31, **31**, **34**, **36**, **65**, 68, 93-4, 94, **110**, **113**, 114, **132**, **140**, **144**, 148-9, **155**, **159**, 168
Tonypandy 19, 132, 144
tools 52-3, 65, 71, 91, 100, 102, 137
    leaving underground 49, 51, 57
    locking bars 57, 102
towns 78, 97, 142, 148-50, 157
train transport 47, 50, 63, 81, 91, 125, 148, 159
trams **10**, 49, 50, 53-4, **57**, 62, **65**, 72, 74, 99, 100, 136, 137, **149**, **155**
    derailment 54, 125
    damage 56-7,
    man-riding 85
Traprain, Viscount (Robert Balfour) 150
Trealaw 6, 18, **18**, 19, 28, **34**, 35, **67**, 118, **139**, **140**
Treharris 8
Tressell, Robert 77
Tudor Hart, Edith 15-16, 25
Tylorstown 17, 31, 35, **86**, **113**, 124n, **144**
unemployment 16, 96, 156, 158
union 9, 12, 32, 116, 120, 148
    ballots 63, 81
    meetings 32, 64, 73, 76, 80-1, 127-8
    committees 64, 72, 76, 91-3, 128
    officials 9, 60, 64, 72-4, 82, 93, 98-9, 120, 156
    leadership 12, 32, 73, 75, 86

dues 32, 75
    pay advice 89, 138
van Gogh, Vincent 15
ventilation 52, 55, 69-70, 121, 130-1
wages. *See* pay
walking to work 10, 33, 81
washing, laundry **23**, 29, 47, 63, 67, **110**, **113**, 139, **157**
waterfalls 10, 12, 59, 94
Welfare halls, libraries 9, 33, 66, 68, 122-2
Welsh Labour Outlook 11
Welsh language 9, 51, 56n, 84, 116, 142, 145, 147, 150
Welsh shawl **37**, 40, **67**, 68, 112
Whit Monday 37, 145
Williams, Chris 8, 42, 43
Williams, Crush (co-worker) 27, 83, 97, 124-5, 133, 134, 143, 147, 150, 152-4, 159
women in mining communities 36-40, 91, 112
    portraits **37**, 43, **96**, **109**
    and babies **37**, **67**, 68
    baths 139, 141
    as carers 28, 120, 137, 147
    housework **13**, 29, 37, 74, 96, 106, 109, 112, 139, 144
    widowhood 80, 116, 148
women's employment 16, 37, 56, 85, 90, 96, 149, 157
Woolf, Virginia 25
Woolworth's 149-50
workmen's committee 50, 63, 76, 91-2, 98
X-rays 28, 60, 95, 157
Ynys-gron (BLC's farm) 13, 168
Zobole, Ernest 25
Zola, Emile 77

Modern Wales by Parthian Books

The Modern Wales Series, edited by Dai Smith and supported by the Rhys Davies Trust, was launched in 2017. The Series offers an extensive list of biography, memoir, history and politics which reflect and analyse the development of Wales as a modernised society into contemporary times. It engages widely across places and people, encompasses imagery and the construction of iconography, dissects historiography and recounts plain stories, all in order to elucidate the kaleidoscopic pattern which has shaped and changed the complex culture and society of Wales and the Welsh.

The inaugural titles in the Series were *To Hear the Skylark's Song*, a haunting memoir of growing up in Aberfan by Huw Lewis, and Joe England's panoramic *Merthyr: The Crucible of Modern Wales*. The impressive list has continued with Angela John's *Rocking the Boat*, essays on Welsh women who pioneered the universal fight for equality and Daryl Leeworthy's landmark overview *Labour Country*, on the struggle through radical action and social democratic politics to ground Wales in the civics of common ownership. Myths and misapprehension, whether naïve or calculated, have been ruthlessly filleted in Martin Johnes' startling *Wales: England's Colony?* and a clutch of biographical studies will reintroduce us to the once seminal, now neglected, figures of Cyril Lakin, Minnie Pallister and Gwyn Thomas, whilst Meic Stephens' *Rhys Davies: A Writer's Life* and Dai Smith's *Raymond Williams: A Warrior's Tale* form part of an associated back catalogue from Parthian.

the RHYS DAVIES TRUST

PARTHIAN

WALES: ENGLAND'S COLONY?
Martin Johnes

From the very beginnings of Wales, its people have defined themselves against their large neighbour. This book tells the fascinating story of an uneasy and unequal relationship between two nations living side-by-side.

PB / £8.99
978-1-912681-41-9

RHYS DAVIES: A WRITER'S LIFE
Meic Stephens

Rhys Davies (1901-78) was among the most dedicated, prolific and accomplished of Welsh prose writers. This is his first full biography.

'This is a delightful book, which is itself a social history in its own right, and funny.' – The Spectator

PB / £11.99
978-1-912109-96-8

MERTHYR, THE CRUCIBLE OF MODERN WALES
Joe England

Merthyr Tydfil was the town where the future of a country was forged: a thriving, struggling surge of people, industry, democracy and ideas. This book assesses an epic history of Merthyr from 1760 to 1912 through the focus of a fresh and thoroughly convincing perspective.

PB / £18.99
978-1-913640-05-7

Bert and his son Peter in January 1941 looking over the Neath valley in a photograph taken by Bert Hardy for *Picture Post*. The view is eastward from the hillside above their then home, Oak Lodge. The conical tip of Aberpergwm Colliery squats on the valley floor next to the tall chimney of the colliery power station. Cwmgwrach is on the right of the photograph. Shortly after this, the family moved to Ynys-gron, a farm on the far side of the valley (near the puff of smoke).